Annals of an
Abiding Liberal

JOHN KENNETH GALBRAITH

ANNALS

of an

ABIDING

LIBERAL

Edited by Andrea D. Williams

A MERIDIAN BOOK

NEW AMERICAN LIBRARY

TIMES·MIRROR

NEW YORK AND SCARBOROUGH, ONTARIO

Material from this book first appeared, often in somewhat different form, in
various publications and is here reproduced by permission:
"The Valid Image of the Modern Economy" under the title "The Bimodal
Image of the Modern Economy: Remarks upon Receipt of the Veblen-Commons
Award" in *Journal of Economic Issues,* June 1977; "Economists and the Eco-
nomics of Professional Contentment" under the title "The Trouble with Economists" in *The New Republic,* January 4, 1978; "The Higher Economic Purpose
of Women" under the title "Women as Economic Interests" in *MS Magazine,*
May 1974; "The Conservative Majority Syndrome" under the title "The Con-
servative-Majority Fallacy" in *New York* magazine, December 22, 1975; "The
Multinational Corporation: How to Put Your Worst Foot Forward or in Your
Mouth" under the title "The Defense of the Multinational Company" in *Har-
vard Business Review,* March-April 1978; "What Comes After General Motors"
in *The New Republic,* November 2, 1974; "The Founding Faith: Adam Smith's
Wealth of Nations" under the title "Scotland's Greatest Son" in *Horizon,* Summer
1974; "Defenders of the Faith, I: William Simon" under the title "Rival Seekers
After Truth" in *The Washington Post,* May 14, 1978; "Defenders of the Faith,
II: Irving Kristol" under the title "A Hard Case" in *The New York Review of
Books,* April 20, 1978; "Defenders of the Faith, III: Wright and Slick" under the
title "The Assault on Private Enterprise" in *The New York Times,* September
15, 1974; "Who Was Thorstein Veblen?" as the introduction to *The Theory of
the Leisure Class* by Thorstein Veblen (Houghton Mifflin, 1973); "A Note on the
Psychopathology of the Very Affluent" under the title "Neuroses of the Rich"
in *Playboy,* February 1974; "My Forty Years with the FBI" in Esquire, October
1977; "The North Dakota Plan" in *The Atlantic Monthly,* August 1978; "Germany; July 20, 1944" under the title "Hitler: Hard to Resist" in *The New York
Review of Books,* September 15, 1977; "The Indian-Pacific Train" under the title
"Across Australia by Train" in *Travel and Leisure,* September 1974; "Seven
Wonders" under the title "Seven Wonders of the World" in *The New York
Times,* November 27, 1977; "Circumnavigation 1978" under the title "The Think-

THE FOLLOWING PAGE CONSTITUTES AN EXTENSION
OF THIS COPYRIGHT PAGE.

ing Man's Trip Around the World" in *Esquire,* February 27, 1979; "Evelyn Waugh" under the title *"The Diaries of Evelyn Waugh"* in *The New Republic,* November 20, 1976; "Anthony Trollope" under the title "Political Novels Past and Present" in *The New York Times,* September 12, 1976; "Writing and Typing" under the title "Writing, Typing, and Economics" in *The Atlantic Monthly,* March 1978; "John Bartlow Martin and Adlai Stevenson" under the title "Adlai Stevenson of Illinois" in *The New York Times,* March 7, 1976; "Last Word on the Hiss Case?" under the title "Alger Hiss and Liberal Anxiety" in *The Atlantic Monthly,* May 1978; "Bernard Cornfeld: Benefactor" under the title "Do You Sincerely Want to Be Rich?" in *The Washington Post,* August 15, 1971; "Robert Vesco: Swindler" under the title "Vesco and the Joy of Swindling" in *New York* magazine, November 18, 1974; "Should Stealing from the Rich Be Punished?" under the title "Crime and No Punishment" in *Esquire,* December 1977; "The Global Strategic Mind" under the title "The Strategic Mind" in *The New York Review of Books,* October 12, 1978; "John Dean, Ambition and The White House" under the title "John Dean's Total Recall of Blind Ambition" in *New York* magazine, November 8, 1976; *"RN: The Memoirs of Richard Nixon"* under the title "The Good Old Days" in *The New York Review of Books,* June 29, 1978; "Power and the Useful Economist" in *American Economic Review,* March 1973.

Library of Congress Cataloging in Publication Data

Galbraith, John Kenneth, 1908–
   Annals of an abiding liberal.

   "A Meridian book."
   Reprint of the ed. published by Houghton Mifflin, Boston.
   Includes bibliographical references and index.
   1. Galbraith, John Kenneth, 1908–   —Addresses, essays, lectures. 2. Economist—United States— Biography—Addresses, essays, lectures. 3. Economics— Addresses, essays, lectures. I. Williams, Andrea D. II. Title.
[HB119.G33A32   1980]   330'.092'4 [B]   80-36825
ISBN 0-452-00544-2

This is an authorized reprint of a hardcover edition published by
Houghton Mifflin Company

MERIDIAN TRADEMARK REG. U.S. PAT. OFF. AND FOREIGN COUNTRIES
REGISTERED TRADEMARK—MARCA REGISTRADA
HECHO EN FORGE VILLAGE, MASS., U.S.A.

SIGNET, SIGNET CLASSICS, MENTOR, PLUME, MERIDIAN and NAL BOOKS are published *in the United States* by
The New American Library, Inc.,
1633 Broadway, New York, New York 10019,
in Canada by The New American Library of Canada Limited,
81 Mack Avenue, Scarborough, Ontario M1L 1M8

First Meridian Printing, October 1980

1 2 3 4 5 6 7 8 9

PRINTED IN THE UNITED STATES OF AMERICA

For Lois and Tom Eliot

Foreword

THE ESSAYS that make up this book have one major connecting link, which is that they are by the same author. Beyond that, they reflect my interests of the last eight or nine years and my feeling that some of the things that I have labored to make worth publishing once should be worth publishing again. The vanity of authors is a precious thing.

The first group of essays is on economic policy and economic affairs. All of them confront in some measure the alleged conservative trend of our time, the feeling that the less privileged and the poor will be better off if they are required to shift for themselves. Thus my title. I confess, however, that I do not think the movement to the right has been as massive or violent as is commonly proclaimed. It is hard to believe that the American people have suddenly rejected the practical compassion that has served them so well for so long. And as I argue in one of these pieces, the perception of a conservative movement is just as effective as the reality for moving the more vulnerable of legislators — or Presidents.

More specifically, this first section assembles my thoughts on economic affairs since I published *Economics and the Public Purpose* in 1973. The juxtaposition of the few vast enterprises and the many small — the basic characteristic of modern industrial structure — and its consequences are the focus of the first essay. This and the modestly more technical piece in the appendix — the latter my presidential communication to the American Economic Association — are the best summaries I can offer of my present ideas on the larger issues of economic organization, including its predilection for inconveniently rapid change. Another essay tells of what happens when economists resist change — of the results for the politicians they advise and the people for whom they prescribe when their minds remain with the past and their recommendations with the reputable applause. Not all of my discontent, however, is with the conservative stereotypes. The liberal's preoccupation with the multinational corporation has always seemed to me an intellectual blind alley. The essay here on this subject was published in the same week in the *Harvard Business Review* and the *New Statesman* in London. The sound and careful readers of the first journal remained calm; those of the second reacted with much well-reasoned abuse. In later essays I go on to other canons of the liberal faith and to contemplate those who, with various shadings of indignation, vehemence and anger, are moved to dissent.

The next section of the book is on personal history very broadly defined. (The pieces therein pave the way for a much more ambitious effort of two or three years hence). One of these essays — my history according to the FBI — was, by a wide margin, the most expensively researched enterprise with which I have ever been associated. The total cost must, with overhead, have run into hundreds of thousands of dollars. It is appropriate that I give the product of all this work to the public, for it was the taxpayers who paid. Several of the other pieces in this group tell of travel. Like many other college professors, I have frequently substituted movement for thought, although some of this has been to appease my wife. She regularly expresses regret over not yet having been to central Greenland; she has been everywhere else.

It is hard, maybe unnatural, for anyone who writes not to be interested in writers. In the section on the arts I have dealt with several of my heroes. But the concept of art is, fortunately, an infinitely elastic one. I have always been fascinated by the various forms of financial prestidigitation and its more inspired practitioners. I even regret, as I here confess, that a confidence man so accomplished that he can play successfully on the avarice and innocence of great bankers — George Moore and Walter Wriston of Citibank, for example — should be sent to jail. I am less fascinated, on the whole, by political thimblerigging of the Nixon type. But its literature, so abundant in the seventies, has also its peculiar interest.

The subject matter of all of these essays is mine; there are some things on which even the most self-effacing author can insist. But that they can be read without puzzlement as to meaning or outrage as to syntax, spelling, mistaken fact or lack of elementary good taste is entirely owing to my friend, partner and editor Andrea Williams.

JOHN KENNETH GALBRAITH

Cambridge, Massachusetts

Contents

II. PERSONAL HISTORY

III. THE ARTS AND . . .

IV. . . . THE DUBIOUS ARTS

I
Economic Affairs

The Valid Image of
the Modern Economy

This article is based on my formal acceptance speech given on receiving the Veblen-Commons Award from the Association for Evolutionary Economics in 1976. A more technical and in some respects more precise statement of this theme was in my presidential address to the American Economic Association in 1972. The latter, I am not quite alone in believing, is the best short account of my general economic position. However, it could, here at the outset, discourage the interested but professionally unconditioned reader, so, fearing this, I have put it into an appendix later in this book.[1]

I AM HERE CONCERNED to see if I can provide a comprehensive and integrated view of the principal problems of economic management in our time. In doing so, I shall offer an alternative picture of the structure of modern economic society. This will compress into brief, and, I trust, sharp form without obscuring detail what I have hitherto written about at much greater length.[2] Finally, I shall attempt to apply this model to some contemporary problems.

In considering the image of modern industrial society, one must have clearly in mind two factors that act strongly and persistently to distort the economist's view of that reality.

The first of these distorting factors is the very great inclina-

---

[1] "Power and the Useful Economist," p. 353.
[2] In *Economics and the Public Purpose* (Boston: Houghton Mifflin, 1973).

tion to think of the ultimate subject matter with which we deal in static terms. Physics, chemistry and geology deal with an unchanging subject matter. What is known and taught about them changes only as information is added or interpretation is revised. They are, all agree, sciences. It is the great desire of nearly all economists to see their subject as a science too. Accordingly, and without much thought, they hold that its matter is also fixed. The business firm, the market, the behavior of the consumer, like the oxygen molecule or the geologist's granite, are given. Economists are avid searchers for new information, eager in their discussion of the conclusions to be drawn. But nearly all of this information is then fitted into a fixed, unchanging view of the role of business firms, markets, labor relations, consumer behavior, and the economic role of the government. It is not an accident that economists who see their subject in evolutionary terms are a minority in the profession.

This is not a small methodological point. You will not doubt its importance if, in fact, the institutions with which economics deals are not stable, if they are subject to change. In truth, they *are,* and the first step toward a more valid perception of economic society and its problems is an appreciation of the very high rate of movement that has been occurring in basic economic institutions. The business corporation is the greatest of the forces for such change. In consequence of the movement it initiates, there has been a rapid alteration in the nature of the labor market and of trade union organization. Also in the class structure of modern economic society and in the resulting patterns of consumption. Also in the services and responses of the modern state. The ultimate effect of these changes is, in fact, to make the economic knowledge of one generation obsolete in the next. And also the prescription and policy based on that knowledge.

The second factor that distorts economic understanding is the very great social and political convenience — or so it seems —

of the wrong image of economic society. I can best give substance to this abstraction by proceeding to the structure of the modern industrial economy.

The presently accepted image of this economy is, of course, of numerous entrepreneurial firms distributed as between consumer- and producer-goods industries, all subordinate to their market and thus, ultimately, to the instruction of the consumer. Being numerous, the firms are competitive; any tendency to overprice products by one firm is corrected by the undercutting of a competitor. A similar corrective tendency operates, if less perfectly, in the purchase of materials and labor. Being entrepreneurial, the firm has a simple internal structure. Authority, power within the firm, lies with the entrepreneur, on whom, overwhelmingly, achievement depends. The entrepreneur being the owner, the partial owner or the direct instrument of the owner, the motivation is also simple and straightforward. It is to maximize return.

To say that the firm is subordinate to the market is to say that it submits to prices that it does not control and that it submits, ultimately, to the will of the consumer. Decision originates with the consumer, and this decision, expressed through the market, is sovereign. If the consumer has sovereign power, the firm cannot have any important power at all in the market; there cannot be two possessors of sovereign power. The business firm is also, by assumption rather than by evidence, without organic power in the state.

In one exception, the firm has influence over prices and output; that is the case of monopoly or oligopoly, or their counterparts, in the purchase of materials and components, products for resale or labor. But monopoly — the control of prices and production in an industry by one firm — and oligopoly — control by a few firms — are never the rule in this image; they are always the exception. They are imperfections in the system. The use of the word imperfection, which is the standard reference to monopoly and oligopoly, affirms that these are departures from the general competitive rule.

To any economist the broad image of economic society that I have just sketched will not seem replete with novelty. It is also admirable proof of the resistance of the subject to change. In the last hundred years the notion of oligopoly has been added to that of monopoly, and the notion of monopoly has been widened to include partial monopoly in brands, services or the like — monopolistic competition. On occasion, there is now in basic economic instruction some bow to the managerial as distinct from the entrepreneurial character of the modern great corporation. Otherwise the basic structure — competitive entrepreneurial firms, the supremacy of the market, the flawing exception of monopoly — is not very different in the modern textbook from that described in Alfred Marshall's *Principles of Economics,* which was first published in the year 1890. Anyone not deeply conditioned by conventional economic instruction must wonder, as he or she reflects on the extent of economic change in our time, if so static a theory of basic economic arrangements can be valid. It is right to do so.

The image is not valid. But it does contribute both to the tranquillity of the economist's existence and to the social and political convenience of modern corporate enterprise.

The service of the accepted image of economic life to the political needs of the business firm — the large corporation in particular — is, in fact, breathtaking. Broadly speaking, it removes from the corporation all power to do wrong and leaves with it only the power to do right.

Are its prices too high? The corporation is blameless. Prices are set by the market. Are profits unseemly? They too are determined by the market. Are products deficient in safety, durability, design, usefulness? They only reflect the will of the sovereign consumer. The function of the firm is not to interpose its judgment, only to accept that of the consumer. Is there adverse effect on the environment? If so, it reflects (with some minor effect from external diseconomies) the higher preference of people for the goods being produced as opposed to the protection of air, water or landscape. Is there criticism of the influence of corporations on the state — of the devastating foreign policy of Lockheed in Italy, Japan, Holland? These are aberrations, for an organic relationship between the business firm and the state does not exist.

One sees how great are the political and social advantages of this image of economic life. It is not easy to think of the accepted economics as the handmaiden of politics. Most economists suppress the thought. None should.

However, self-delusion also has its cost — and this is great. Specifically, this image conceals from us the workings of the modern economic system, the reasons for its successes and its failures and the nature of the needed remedial action. Among the victims of this concealment are those most intimately involved — those with the greatest need to understand the correct image — and they are businessmen themselves. And there is a damaging public effect. People cannot accept as valid an image of modern society that makes the great corporation the helpless, passive instrument of market forces and itself a force of minimal influence in the state. This is too deeply at odds with common sense. So they come to believe that there is something intrinsically deceptive about the modern corporation, and perhaps also about the economics that projects the conventional image. Better and safer the truth.

The valid image of the economic system is not, in fact, of a single competitive and entrepreneurial system. It is of a double or bimodal system. The two parts are very different in structure but roughly equal in aggregate product. In the United States, reflecting the force of the corporation for change in the last century, around 1000 to 2000[3] firms contribute about half of all private economic product. In 1967, for example, 200 manu-

[3] The statistical difference between 1000 and 2000 is not, in fact, great, for the contribution of the second thousand is small as compared with that of the first.

facturing corporations (out of 200,000) shipped 42 percent of all manufactured goods by value. Later figures suggest further concentration. Of 13,687 commercial banks in 1971, 50 had 48 percent of all assets; of 1805 life insurance companies, the 50 largest had 82 percent of all assets.[4] Set against this half of the economy is the dispersed sector; depending on what is called a firm, this consists, in the United States, of between 10 and 12 million small businesses — farms, service and professional enterprises, construction firms, artistic enterprises, small traders. They contribute the other half of product. The division in other advanced industrial countries is roughly similar. Thus the valid image of modern economic society is the division of the productive task between a few large firms that are infinitely large and many small firms that are infinitely numerous.

The large corporation differs organically from the small; the burden of proof cannot seem excessive for the individual who asserts that there is a fundamental difference in organization and structure between General Motors, Shell or Volkswagen and the small farm, neighborhood restaurant, cafe or retail flower shop. The coexistence of these two very different structures and the resulting economic behavior are themselves features of the greatest importance. But first a further word on the corporate sector — what I have elsewhere called the planning system.[5]

The most obvious characteristic of the corporate half of the economy is the great size of the participating units. In the

[4] Jonathan R. T. Hughes, *The Governmental Habit: Economic Controls from Colonial Times to the Present* (New York: Basic Books, 1977), p. 203. William Leonard, adjusting for some underreporting — the tendency to assign some manufacturing activities to mining for tax reasons — puts the share of manufacturing employment of the largest 200 corporations at 60 percent in 1974. "Mergers, Industrial Concentration, and Antitrust Policy," *Journal of Economic Issues,* vol. X, no. 2 (June 1976), pp. 354–381.

[5] In *The New Industrial State,* 3rd ed. (Boston: Houghton Mifflin, 1978) and *Economics and the Public Purpose.*

United States a handful of industrial corporations — General Motors, Exxon, Ford, a couple of others at most — have sales equal to all agriculture. Size in turn contributes to the two features of the modern large firm that differentiate it from the entrepreneurial and competitive enterprise and explain its impact on the society. The first of these is its deployment of market and political power. The second — one that is less noticed — is its diffusion of personal power.

The deployment of market and political power is diverse and, except as described in economic instruction, also commonplace. The modern large corporation has extensive influence over its prices and over its costs. It supplies much of its capital from its own earnings. It strongly influences the tastes and behavior of its consumers; even professional economists when looking at television have difficulty concealing from themselves the impact of modern advertising, although many succeed. And it exists in the closest relationship with the modern state.

The government gives the corporation legal existence; establishes the environmental and other parameters within which it functions; monitors the quality and safety of its products and certain of the advertising claims it makes for them; supplies, in the manner of highways to the automobile industry, the services on which sale of its products depends. Also — an increasingly important function — the government is the safety net into which the firm falls in the event of failure. Above a certain size — as the recent history of some large American banks, the eastern railroads in the United States, the Lockheed Corporation, Rolls-Royce, British Leyland, British Chrysler, Krupp and the vast agglomerations of IRI in Italy all show — a very large corporation is no longer allowed to go out of business. The social damage is too great. Modern socialism is extensively the adoption by reluctant governments, socialist and otherwise, of the abandoned offspring of modern capitalism. Being thus so dependent, the corporation must seek power in the state. This power, like that in the market, is not plenary. But its existence can be denied only by those who are trained extensively to ignore it.

As earlier noted, the role of the modern great corporation in diffusing personal power is less celebrated than its deployment of market and public power, but it is not, I believe, less important. In its fully developed form, the corporation, as others have emphasized, removes power from the ownership interest, the traditional locus of capitalist authority. In doing so, it removes it from the representatives of the stockholders — the board of directors. No director of General Motors, Exxon or IBM who is not a member of management — I speak carefully here — has any continuing effective influence on company operations. The ceremony which proclaims that power — usually of aged, occasionally senile men meeting for a couple of hours on complex matters six times a year — is almost wholly implausible except to the participants. Directors do not make decisions; they ratify them. But to remove power from the owners and their alleged representatives — from the capitalists — is only a part of a larger process. That larger process involves extensive diffusion of such power. As power passes from capitalist to management in the large firm, this diffusion occurs in three ways.

First, decisions being numerous and complex, they must be delegated and redelegated, and the decision-making process passes down into the firm. This all recognize to be necessary. Nothing so criticizes an executive as the statement, "He cannot delegate responsibility."

Second, decisions being technically and socially complex, they become the shared responsibility of specialists — engineers, scientists, production men, marketing experts, lawyers, accountants, tax specialists. Power, in other words, passes from individuals to groups — to what I have called the technostructure of the modern corporation.

Finally, where there is no participation in decision, organization takes form to influence it. Thus the trade union. Union power is the natural answer to the power of the corporation. Only in the rarest cases in the developed industrial world is there a large corporation where labor is not organized.

The diffusion of power extends beyond the boundaries of the corporation, for the corporation brings into existence a vast array of supporting professions and services — law firms to advise on, or sometimes bend, the law; accountants to record, and sometimes create, its earnings; universities, colleges and business schools to train its executives and specialists or those who will so pass; dealers to sell its products; repairmen to service the products or advise that they are beyond repair. Marx held that, in its final stages of development, the capitalist firm devoured the small entrepreneur. This may well be true as regards small competitors. But the modern corporation also nurtures and sustains a large penumbra of independent firms. These peripheral groups and firms also assert their right to power. Lawyers and accountants have their special claims on decisions. So do consulting firms and custodians of expert knowledge from the universities. Dealer relations departments exist to consider the rights of those who sell and service the products. All have a claim on power.

We should not test our image of the economic system by its political convenience, or we should not if we are interested in analytically serviceable truths. We should see, instead, whether our image accords with observed circumstance, observed need.

The first test of the system I have just been describing has to do with the foremost problem of our time, the disagreeable and persistent tendency for severe unemployment in the modern industrial society to be combined with severe inflation.

If one accepts the competitive and entrepreneurial image of economic society, this combination does not and cannot occur. There can be inflation. But by conventional macroeconomic monetary and fiscal policy — restricting bank lending and tightening the public budget — the aggregate demand for goods and services in the economy can be reduced. Since, in this image, no firm controls prices, production is affected only as prices fall — that is what brings to the firm the message of declining demand. So, as the first effect, prices will cease to rise, which is to say the inflation will come to an end. Later, as prices and earnings fall, production may be curtailed and there may be unemployment. But unemployment and inflation do not and cannot coexist. One is cured before the other is caused.

Similarly, if there is unemployment, aggregate demand in the economy can be expanded by monetary and fiscal action — more public expenditure, reduced taxes, easier lending and thus more spending from borrowed funds. The initial effect will be more sales, more jobs. Prices may then rise. But, once again, that is because unemployment has been cured or, at a minimum, is by way of being cured.

In the bimodal image of the economy, a combination of inflation and unemployment must be expected at least for so long as fiscal and monetary policy are the sole instruments of economic management. Trade unions, as we have seen, have power over their wages in the corporate sector of the economy. Corporations, having power in their markets, have the ability to offset concessions to trade unions with higher prices. Modern collective bargaining has lost much of its old-fashioned acerbity for a very simple reason: as an alternative to confrontation, unions and management can reach agreement and pass the resulting cost on to the public. Complaints over the cost of wage settlements now rarely come from employers. Almost invariably they come from the government, which is concerned over the inflationary effect, or from the public, which has to pay the higher price.

When this wage-price inflation is attacked by the traditional methods — monetary and budget restraint to reduce demand — prices do not automatically fall. The firm has the power to maintain its prices. The first industrial effect is, instead, on sales, output and employment. And if unions continue to press for higher wages, prices will continue to increase. Only when unemployment is very severe — so severe as to deter the unions from pressing for wage increases and the corporations from exercising their power to raise prices — do the traditional monetary and fiscal measures begin to bite. Meanwhile unemployment and inflation, as in the world today, do coexist.

Before monetary and fiscal policy act on the corporate sector, however, they work on the competitive and entrepreneurial sector of the economy. Here, as before, prices do respond to monetary and fiscal measures to restrain demand. Also in this half of the economy are industries — housing and construction being the notable cases — that exist on borrowed funds, which makes them uniquely vulnerable to monetary action, to restrictions on bank lending. (This vulnerability is in contrast with the position of the large corporation, which has resort to retained earnings for capital and which, in the event of outside need, is a priority customer at the banks.) So, while inflation continues in the corporate half of the economy, there can be falling farm prices and a painful recession in the entrepreneurial and competitive sector. That too accords fully with recent or present circumstances. Beginning in 1974, monetary restriction was brought sharply to bear on the then serious inflation. There followed a serious recession, the worst, in fact, since the Great Depression. Farm prices fell. Housing, where output fell by more than a third, was seriously depressed. Unemployment rose to around 10 percent of the labor force. And industrial prices — those of the corporate sector — kept right on rising.

The practical conclusion is that inflation cannot now be arrested by fiscal and monetary policy alone unless there is willingness to accept a very large amount of unemployment. There remains only one alternative; that is to restrain incomes and prices not by unemployment but by direct intervention — by an incomes and prices policy. Such action is not a substitute for orthodox monetary and fiscal management of demand but an essential supplement to it.

There is a further test here of the validity of the revised

image, for the policies appropriate to it reflect the direction in which most of the industrial countries of the nonsocialist world are moving — against the advice of all the more clamorous voices of conventional economics. In Germany, Austria, Switzerland and Scandinavia wage negotiation is in accordance with an implicit incomes policy that considers the effect of wage concessions on both domestic inflation rates and external competitive position. Britain, a peculiarly resistant case, has, at this writing, a comprehensive incomes and prices policy. France has a more limited one. And the United States government in its guidelines has conceded the need for such a policy and is reluctant only to bring it to effective reality.

The bimodal view also explains the increasingly unequal development of the modern economy and the measures that governments find themselves taking to deal with it. The corporate half of the economy combines advanced organization, high technical skills and relatively ample capital with the ability to persuade the consumer and the state as to their need for its products. In consequence, in all industrial countries, automobiles, lethal weapons, household appliances, pharmaceuticals, alcohol, tobacco and cosmetics are amply supplied. The very notion of shortage, inadequacy, in these commodities would strike all as distinctly odd. The contemporary experience with oil shortages is deeply traumatic. But in the competitive entrepreneurial sector, where organization, technology, capital and persuasion are less available or absent, inadequacy is assumed. Housing, health care, numerous consumer services and, on occasion, the food supply are a source of complaint or anxiety in all of the developed countries. All governments find themselves seeking ways to compensate for the inadequacies of private enterprise in this half of the bimodal economy. The conventional economics has only one explanation for this unequal development: it reflects consumer choice, which is to say that the consumer is unaware of his — or her — needs. Where housing, health care and food are concerned, this is hard to believe.

The bimodal image of the economy serves also our understanding of inequality of opportunity and reward in the modern economy and its consequences. In the conventional image of the economy, inequality is the result of differences in talent, luck or choice of ancestors. But between occupations it is constantly being remedied by movement from lower- to higher-income jobs. If this remedy is to work, people must, of course, be able to move.

The corporate sector of the economy deserves more approval than it receives for the income it provides. In the United States it is doubtful if any union member with full-time employment in this sector falls below the poverty line. But there are grave barriers to movement into this area. In particular, so long as inflation is the chief problem and monetary and fiscal policy are the remedies, there will be unemployment in this sector — either chronic or recurrent. If there is unemployment, there obviously cannot be easy movement of new workers into its higher paid employment. The old unemployed have first chance.

In the entrepreneurial part of the economy, by contrast, employment can often be found either by taking a lower self-employment return or possibly low pay in an industry that has no union. There is, accordingly, a continuing source of inequality between the two parts of the economy derived from the occluded movement between them. We have here another reason for forgoing exclusive reliance on monetary and fiscal policy for controlling inflation. The resulting unemployment is also a source of occluded movement and thus of further inequality as between the different sectors of the economy.

If fiscal and monetary policy alone are used to control inflation in the modern economy, it will be controlled only by

creating unemployment. There must, as noted, be enough unemployment to require unions to forgo added wage claims and
to cause consumers (and corporations) to resist price increases.
Or there must be an incomes and prices policy. No two countries are likely to resolve this problem with the same choice of
measures. In recent years Switzerland, Austria and the German
Federal Republic have had low rates of inflation. Something
must be attributed to economic wisdom. But more must be
attributed to governments that have a history of concern for
inflation and trade unions that are cautious about pressing
inflationary wage claims. And something must also be attributed to the policy of balancing out the labor force with im-
ported labor. It eases social tension if some of the unemployed,
when not needed, are in Italy, Spain, Turkey or Yugoslavia.
In Britain and the United States the reserve unemployed are
within the country itself.

For the above and yet other reasons different countries solve
the unemployment-inflation problem in different ways and with
differing degrees of success. The result is widely differing degrees of inflation in the several industrial countries. With different inflation rates there will be, it is certain, compensating
movements in exchange rates, and there is no formula for
international currency stabilization that will produce stability
in the international exchanges in face of these widely varying
rates of internal inflation. This is something to remember
whenever one hears that central bankers and other monetary
experts are meeting on international currency reform. In the
absence of broadly coordinate policies to control domestic inflation, there can be no international exchange stabilization
that has any hope of being permanent. Promises to the contrary
are a fraud.

There are further tests of the image of the economy I am
here describing. Let me conclude by combining several into
one. In removing power from owners, diffusing it through the

technostructure and accepting and even nurturing the organized response of workers, the modern corporation does more than diffuse power. It takes a long step, if not toward a classless society, at least toward one in which class lines are extensively blurred. This, in turn, has a major effect on consumption patterns. Specifically, there is no longer in the corporate sector of the economy full acceptance by any group that it was meant by the nature of its occupation to consume less. And this acceptance will continue to erode. The pressure so exerted both for private goods and services and the requisite wages is one source of inflation. Pressure for such public services as education, health care and public transportation is another source. The thrusts for more private income and consumption and for more public goods and services have, we see, the same sources and can be equally strong. They are associated with the power — the power diffused by the corporation — to make the claim effective. In consequence, to cut consumption of private goods through taxes or for that matter through an incomes policy or to cut the consumption of public goods through reduced public outlays is very difficult. The bimodal image of economic society helps explain the new budget pressures with their inflationary effect as well as the new sources of inflation in the wage-price spiral. And it tells us also why control is politically and socially so difficult.

The business units in the corporate sector of the economy, becoming large, become international. The modern corporation internationalizes its income and wage standards as entrepreneurial industry never did. It also creates an international civil service — men who, like the servants of the Holy Church, are at home in all lands, who differ only in owing their ultimate allegiance not to Rome but to IBM. The international corporation defends relative freedom from tariff barriers and other constraints on trade. That is because competition is rarely cutthroat between large firms; they are restrained by oligopolistic convention. And international competition is never serious if you own the international competitor. It was the growth of the corporate sector of the modern economy that made possible the Common Market — made it necessary, perhaps, because intra-European trade barriers had become only a nuisance for the large corporation. Agriculture and other entrepreneurial enterprises have not changed their attitude on international trade. Their instinct is still protective. Farmers and other small producers would never have brought the EEC into existence. They are the source of at least 90 percent of its problems. Again the bimodal image fits the history.[6]

Finally, the image of the economy here offered explains the new tensions in the relationship between economic institutions and the state. The competitive and entrepreneurial firm seeks services from the state; seeks protection from competition, as just noted; is subject to regulation; pays taxes. This is a familiar and limited relationship. This firm never, by itself, competes with the state in the exercise of power. The modern large corporation, on the other hand, has a far wider range of requirements from the state. It also brings its power directly to bear on the instrumentalities of the state — both the bureaucracy and the legislature. Its needs, since they are put forward by the technostructure — an influential and articulate sector of the population — have a way of becoming public policy. Americans have recently had a substantial education in the way the financial resources of the corporation have been deployed for the purchase of politicians and political influence. And in numerous matters the corporation exercises power of a purely public nature. In recent years the aircraft companies have had more success, of a sort, in the making and unmaking of foreign politicians and governments than has the CIA. No one doubts that the oil companies conduct a policy in the Middle East that sometimes supersedes that of the Department of State. A

---

[6] I return to this theme in the later essay, "The Multinational Corporation: How to Put Your Worst Foot Forward or in Your Mouth."

good many people believe that General Motors has had considerably more to do with setting policy on mass transportation in recent times than has the United States government.

These tensions are a great and important fact of life. As with inflation and unemployment, unequal development and inequality, we presently deal with them in the industrial countries by resort to an image of industrial society which holds that they do not exist. Or which holds that they are aberrations *sui generis*. This is unconvincing to the average citizen who, unlike the more acquiescent economist, is untrained in illusion. It precludes effective diagnosis and effective remedial action. It is safer and wiser as well as intellectually more rewarding to accept the reality.

2

Economists and the Economics
of Professional Contentment

BY THE TIME of the Great Depression of the nineteen-thirties economics had become a subject of respected instruction, research and public guidance in the United States and was an academic discipline of no slight prestige. Harvard, Columbia, Yale, Chicago, Princeton, California and Wisconsin were major centers of such effort. F. W. Taussig, the noted Harvard teacher, tariff-maker and wartime price-fixer; Joseph Schumpeter, who came to Harvard in these years from Germany; Wesley C. Mitchell at Columbia University; Fred R. Fairchild at Yale, whose textbook with Furniss and Buck instructed and depressed many undergraduate generations; all had reputations of national reach. In universities and colleges around the country there were others of only slightly less esteem. In other countries, especially in Britain and Germany, economists enjoyed similar fame.

The Great Depression, beginning in the autumn of 1929 and continuing for ten years until washed out by military expenditures in 1940–41, was, rivaled only by the Civil War, the most traumatic event in the American experience. It lingers more strongly than any other in the social memory of Americans. When the economic prospect is uncertain, men and women still ask in fear, "Will there be another depression?" In those years output — Gross National Product — fell by a third. Farmers, still numerous and vocal, went bankrupt and were dispossessed by the tens of thousands. By the mid-thirties the farm debt exceeded the assets, at current values, of all farms. American agriculture was, literally, insolvent. Unemployment rose to around a quarter of the labor force. There was no unemployment insurance. Until 1933, there was no effort to enhance job opportunity and until then no organized assistance to the impoverished. As now in relation to the urban crisis, men of high reputation took their stand on deprivation on principle; far better suffering and despair than any impairment of the rule of local responsibility for local problems.

Nor was the misfortune only that of the poor. By the end of March 1933, 9000 banks had failed and around 100,000 other commercial and industrial enterprises. By the mid-thirties bankruptcy was no longer the misfortune of the weak and the small; numerous of the large banks, utilities and railroads were in receivership or in peril. The president of the New York Federal Reserve Bank had observed that while "[the effects of the failure of] the small banks in the community could be isolated," danger to the big New York banks had to be taken seriously.

The great economists, with a few exceptions, reacted to these misfortunes with professional detachment and calm. Called on for advice, as they were, most warned of the dangers of "untried experiment," experiment usually being of such character. Or they stressed the danger of inflation. The United States had had a serious inflation in World War I and following; prices had approximately doubled. In Germany and Austria and elsewhere in Central Europe, the purchasing power of currencies had totally collapsed, with maximum impact on a generation of scholars, many of whom later migrated to the United States. In the manner made famous by generals, the great economists showed a marked affinity for fighting their past wars and did so through a deflation that, between 1929 and 1933, brought the wholesale price index down by more than a third.

There was also, in these years, notable stress on the importance of patience as a therapy — a treatment that is believed to be easier with academic tenure on a regular income. Especially powerful were the warnings from Joseph Schumpeter of Harvard and Lionel Robbins of the London School of Economics. They affirmed that depressions ended only when they have corrected the maladjustments and extruded the poisons by which they were caused.

On occasion there was organized effort to resist action or support inaction. In the autumn of 1933, many of the more prestigious members of the profession banded together under Edwin W. Kemmerer of Princeton to oppose monetary and fiscal experiment. A year or so later several Harvard economists collaborated in an astringent attack on the economic experimentalism of the New Deal. This was especially notable, for the effort involved not only the older professors but also younger members of the faculty who were still actively pursuing tenure. The most common mood, however, was one of judicious scholarly contentment. Professors remained concerned with accustomed teaching, research and writing in economic theory, money and banking, statistics and economic history. At the great universities there were few tense seminars on the causes of the Depression or its cure. Little such discussion was reflected in the professional journals. Sometime in 1934 or 1935, Roosevelt returned to Harvard to dine with his sons and their friends at their undergraduate club. As he passed along the street, he was greeted by students with shouts of "Fire Tugwell!" This reflected the dominant lesson of their practical economic instruction; among the economists in the administration, Rexford Guy Tugwell was the most notable activist and thus the most widely criticized.

There were exceptions to this mood — the professional eccentrics and deviants. Many university campuses had one or two — men or the occasional woman who kept asking what might be done nationally, locally or on specific matters to alleviate the suffering. Agricultural and labor economists, living close to their clientele, had a special tendency to be involved, and at considerable cost to their scholarly reputations. Academically speaking, they were always second-class citizens. At the University of Wisconsin, under John R. Commons, there was a widely remarked concern with such issues as unemployment insurance, utility rate regulation and taxation. Wisconsin was thought by many a peculiar place.

Elsewhere two men were outspoken in advocacy of their preferred remedies for the Depression. One of these was Irving Fisher of Yale; the other, of Cambridge and various London preoccupations, was John Maynard Keynes. Fisher urged deliberate expansion of the money supply through reduction of the gold content of the dollar. Keynes urged energetic expenditure by governments from borrowed funds. Both men were regarded with deep distaste by the economists of established reputation. The *New York Times* in 1933 thought it "should hardly be necessary to say" that the ideas of Fisher and Keynes "have been long before the public, and that both have been rejected by the large consensus of economic and financial judgment." Keynes and Fisher, with Commons, are among the few economists from the time who are not forgotten. And the veil is kind, for, with the passing years, those who continued to sit stolidly on their prestige paid heavily for their contentment. Their comfortable negativism, so far as it is remembered, is treated with amused contempt in the intellectual history of these years.

I recur to this history because detachment and contentment are again the tendency of economists in our time. Politicians and the public, and not least economists, should know that it is a normal cyclical phenomenon in our profession. What is understood may not be more easily forgiven, but there is a chance that it will be more promptly remedied.

The contentment of the early Depression years was followed in the late thirties, in the war years and thereafter, including through the Eisenhower years, by a period of excitement and active innovation in economic policy. This was overwhelmingly in reaction to the increasingly visible acquiescence and ineptitude of the great men of the past. The polar figure was, of course, Keynes; the unifying idea was the belief that something could be done and had to be done by government to lift and sustain the level of output and employment in the modern capitalist economy. There was much here to be discussed. And the discussion was further sustained by the myriad of new economic tasks occasioned by the war and by the development and use during these years of the National Accounts — the incorporation into professional thought of National Income, Gross National Product and aggregates of consumer and business spending and saving. Almost simultaneously there was a breakthrough in the ideas as to what should be done to overcome unemployment and depression and in the measurement thereof. Again, however, the very great men of the profession, with such rare exceptions as the late Alvin Hansen of Harvard, continued aloof. Men of lesser reputation with less reputation to lose made the Keynesian Revolution.

The tactically careful and self-regarding had reason for caution. Those actively associated with the new policies in government were, like Tugwell, subject to severe rebuke. The most general complaint was from businessmen; it was that economists were impairing or destroying confidence. Some reactions were very specific; when the Employment Act of 1946 was under consideration, the National Association of Manufacturers, through Donaldson Brown, a General Motors executive, filed a brief holding that it would enhance government control, destroy private enterprise, unduly increase the powers of the Federal Executive, legalize federal spending and pump-priming, bring socialism, be unworkable, impractical and promise too much. There were also other defects. No prudent scholar would wish to be associated with such destructive activity.

But the heaviest charge was the more general effect on confidence. The American business psyche is an acutely vulnerable thing; it associates all change with perverse ideological intent. No large group of similar size and fortune in history has ever been so insecure as the American business executives. The only form of reassurance that serves them is either lower taxes or inaction, and these are required and expected no matter who wins elections. However, in the thirties and forties and continuing in the fifties and into the sixties, the results of inaction seemed intolerable, and not least for business itself. So the onslaught was faced. Economists accepted their controversial role. And, in time, the more secure and mentally accessible executives came to accept the Keynesian rescue.

With the innovative and combative mood of economists went nearly a quarter century of economic success. At the end of 1968, the President's Council of Economic Advisers congratulated itself on its accomplishments. The language, which then did not seem remarkable, is now rather wonderful to recall:

> The Nation is now in its 95th month of continuous economic advance. Both in strength and length, this prosperity is without parallel in our history. We have steered clear of the business-cycle recessions which for generations derailed us repeatedly from the path of growth and progress . . . No longer do we view our economic life as a relentless tide of ups and downs. No longer do we fear that automation and technical progress will rob workers of jobs rather than help us to achieve greater abundance. No longer do we consider poverty and unemployment permanent landmarks on our economic scene . . .
>
> Ever since the historic passage of the Employment Act in 1946, economic policies have responded to the fire alarm of recession and boom. In the 1960's, we have adopted a new strategy aimed at fire prevention — sustaining prosperity and heading off recession or serious inflation before they could take hold . . .
>
> Meanwhile, a solid foundation has been built for continued growth in the years ahead.[1]

These words will suggest what all should have feared — that economics had once again settled into a mood of self-congratu-

1 *Economic Report of the President, 1969*, pp. 4–5.

lation with its associated contentment. And this was happening just as the Keynesian measures, once so wonderful, were showing by the clearest of all possible evidence that they produced not inflation, not unemployment, but an unyielding combination of the two. The evidence was in the plain history of the next ten years. The only relief from this combination would be for a brief period preceding the presidential election in 1972, a breathing spell which reputable economists have dismissed as the purchase of an election at the price of far worse troubles to come. Not again would economists speak of strategies for "sustaining prosperity and heading off recession or serious inflation before they could take hold." Or say that poverty and unemployment were not "permanent landmarks on our economic scene."

In 1977, when he assumed office, President Jimmy Carter accorded effective Cabinet rank to five economists of the highest professional qualification. A Ph.D. in economics replaced a law degree as the basic license for practicing the science and art of public administration. In the next two years unemployment receded somewhat, but for the disadvantaged it remained very high. Inflation got much worse. Plainly these talented men and women came to office at a dismal moment in the history of economics.

They were, in fact, caught in another of the great downswings which, as in the Great Depression, render the profession innocuous or worse. In the universities and research institutes the mood was not totally bland. There was an active academic discussion of tax incentives and penalties for holding the line on prices and wages, a policy that would require the government to proclaim and, in effect, enforce standards for permissible increases. This discussion led on, inevitably, to a consideration of alternatives, including guidelines and controls. Numerous younger economists and some older ones were not in doubt as to the futility of the accepted approach to inflation and unemployment. But the more general response was passive. The small tasks of more pleasant times and their refinement continued to command major attention. Perhaps the accepted design for monetary and fiscal policy did not work. But it could still be taught and also avowed in Washington, for, after all, it once had served. It is essential that economists and noneconomists alike understand the reasons for this recurrent mood. They are deeply ingrained in the sociology of the profession.

The sociology begins with the instinct of economists for applause, an instinct by no means exceptional to our profession. And with the companion wish to avoid rude controversy. These attitudes must then be set against the deeply inconvenient fact that whatever is good for economic performance will always be deeply controversial. And what is most acclaimed for sustaining business confidence — here I speak with caution — will almost always be bad for economic performance. All have heard of contradictions in modern capitalism; all should know that the major immediate contradiction is here. In its great moments economics has understood this contradiction. At the low points in its cycle, as of late, it has not.

In the great creative years during and following World War II, it was taken for granted in Washington that to be effective was to arouse business hostility. Likewise in the universities. Robert M. Hutchins once noted that economics appointments were the most perilous with which, as president of the University of Chicago, he had to deal, and, it may be added, he and his successors in office made selections that were well designed to minimize the risk. When economists avoid controversy and reach for applause, their advice becomes worse than worthless; it becomes affirmatively damaging. It follows that, unless this is clearly understood by economists in responsible positions, their service will also be affirmatively damaging.

These truths — laws I am prepared to call them — must, some will say, be motivated by a deeply antipathetic business sentiment on my part. It is not so. In contrast with many of my liberal friends, I long ago came to terms with the American business system. Liberals, many of them, would break up the large corporations that now, in the number of a thousand or two, account for more than half of all private product in the United States. I would not. Conservative economists would move similarly against the unions. These, like the large corporations, I accept as part of the broad irreversible current of history. I not only wish to see the system survive, but so deep is my effectively conservative commitment that I want it to have the first essential for survival, which is that it work. I am only persuaded that, as experience so well shows, economists have a remarkably plain choice: they can be popular and applauded in the short run and be failures in the long run. Or they can be controversial in the short run and a success in the long run. I pass over a third possibility, which is to be innocuous in both the short run and the long run. That requires no special instruction.

Controversy is inevitable both as regards effective policy to expand the economy and effective action to restrain inflation. All policy to expand the economy will be more successful the more successfully it provides income to those who need it most. The expenditure of that income is then prompt and complete. Such policy also usefully reduces social tension, for, although much effort has been devoted to showing the opposite, income is a remarkably useful antidote to poverty and its associated discomfort. But the policy that professional business spokesmen and the affluent in general will always most approve, the policy that is best for business confidence, will always be that which gives these more fortunate groups the most after-tax income.

The situation is similar where inflation is involved. Any policy that restricts or restrains the freedom to set prices and incomes will be heavily attacked and will be held bad for business confidence. But in a world of large corporations and strong unions no policy that contends successfully with inflation can avoid some restraint on the freedom of business action — on prices charged and incomes paid or received. So a successful anti-inflation policy must also be sharply controversial — and again bad for confidence or what is so described. And the capacity to articulate alarm and to be heard is strongly correlated with business position and income. These voices are wonderfully audible; those of the poor are not. No tendency in modern political economy, as I shall argue elsewhere in these essays, is so powerful as for the voice of the affluent, and that of the business spokesmen in particular, to be mistaken for the voice of the masses.

A commonplace example will illustrate my point. Nothing is more regularly advocated as a support to business confidence than a large horizontal cut in the personal income tax accompanied by an equally substantial reduction in the corporate income tax and a compensatory slash in welfare expenditures. Such action is reliably praised by professional business spokesmen for its motivating effect. In 1978, a tax cut then, as usual, being under discussion, the *Wall Street Journal* put the prospective business response with admirable clarity:

> A general tax cut is well worth trying . . . provided it is not shaped by perverse theories. The key is to let producers keep more of what they produce, and the biggest effect will come from cutting rates where they are the highest. If the tax cuts reduce rather than improve rewards for the economy's most gifted, talented and skilled producers, they will be worse than nothing at all.

Mr. Alan Greenspan, with Mr. William Simon the principal source of President Gerald Ford's economic advice and thus, alas, of his ultimate defeat, went further. He held that a tax cut wholly confined to corporate income would be the very best for business confidence. We are a friendly people. We listen respectfully even to established architects of political disaster.

However, the applause apart, such tax cuts have little positive business effect. They have little motivating effect on the modern organization man — the significant business figure in our time. That is because he is already required by all the weight of the organization ethic to give his best to the business, and he does. Such tax reduction has little immediate effect on

either consumer spending or business investment. If profit
prospects are good, a corporate tax cut is not needed to encourage investment. If they are bad, no tax cut will make them
good. As practical experience with past tax reduction has shown
(and as was duly reported by the Council of Economic Advisers), the initial effect of a cut in personal taxes is overwhelmingly to increase savings. Income so saved does not buy goods,
and modern business investment, some special pleading to the
contrary, proceeds independently of the supply of savings. Only
good economic performance — good employment and demand
— will encourage borrowing of these funds. The reduced welfare spending that would be so much applauded would, of
course, reduce demand.

In contrast, an energetic jobs program to train and employ
the poor, black and young or, in the absence of jobs, provide
the income that, with whatever damage to the soul, does do
something for the body, would have a deeply adverse effect on
business confidence. So likewise expenditures to help make life
in the large cities, if not pleasant, at least safe and tolerable.
So federal spending to reduce the dependence of cities on regressive property taxes, a point affirmed since this was first
written by the California backlash on property taxation reflected in Proposition 13. So, in particular, building and rehabilitation work in the central slums. And the expansive effect
of action along these lines is optimal; money so distributed
is put immediately into circulation by people whose marginal propensity to spend is one. The effect on the performance of a lagging economy is total — almost nothing is lost to
savings.

There is another advantage in such action. As noted, modern
unemployment is highly structured. Shortages of labor in numerous areas and occupations are combined with disastrous
surpluses among the minorities, those without work experience,
among women and those in the urban ghettos away from job
opportunities. The measures just described are by far the best
for reaching those most in need of jobs. Tax reductions, even

if widely distributed among various income groups, add to pressure on labor markets that are already strong. Measures that are directed specifically at the unemployed — targeted, in the offensive current jargon — have their initial effect on labor markets that are weak. They are, in consequence, much less inflationary than action that rewards the affluent.

There is a further inverse correlation between what serves business confidence and what best serves the future of capitalism. None surely can doubt that the long-run future of capitalism will be more secure if the poor, the black and the young have the stake in the system that a steady income provides. There is much talk these days of a taxpayers' revolt. It will not, in the higher-income brackets, be the kind of revolution that involves much raw violence. The revolutionary impulses of David Rockefeller, Walter Wriston and Gabriel Hauge can be contained. One cannot be so sanguine about anger in the slums.

So both control of inflation and the prevention of unemployment involve controversy and are held to impair business confidence. But it is not easy to be overtly in favor of either misfortune. Inflation suggests loose public management; there is ample evidence that people both so regard it and much dislike it. Unemployment has a few more open supporters. The defense of a "natural level" of joblessness unites technical economists who eschew all concern for political consequences with the editors of *Fortune*. And as Robert Lekachman of Lehman College in New York has pointed out, there are many who have noticed that a little fear makes a labor force more productive or acquiescent or both. But no economist in untenured public office can be in favor of unemployment. So, given the need to be against inflation and unemployment and also the need to avoid the controversy and adverse business reaction inherent in all effective remedial action, economists have only one choice. That is some talented form of evasion. This means, in practice, the simulation of action as a substitute for action.

*

In the universities such evasion is relatively easy. As in the
thirties, one can remain with accustomed preoccupations and
ignore inflation or the now associated unemployment as irrelevant to one's particular specialization. A posture of scientific
preoccupation is then adduced to support such retreat. The
scientist, all agree, is an unworldly figure undiverted by practical concerns. So an economist can, in good conscience, reject
the world and be proud of his emancipation from useful matters. In public life the simulation of activity as a substitute for
action involves more varied techniques. These range from
banality to sophisticated fraud.

The most engaging such effort in modern times was President Gerald Ford's invention and distribution of the WIN —
Whip Inflation Now — buttons. It was attractive because it
was transparently honest in intent — in seeking to suggest action without risking any of the pain or controversy that action
requires. Economists in or sometimes out of official office usually prefer more sophisticated designs.

The first and most popular of these designs is to avow that
some as yet uninvoked form of monetary magic will reconcile
stable prices with low unemployment. If Professor Milton
Friedman or anyone else could indeed achieve such a result, it
can hardly be believed that his revelation would have remained
so long unused.

A second evasion involves government regulation. When
economists were assembled by President Ford in 1974 to attack
inflation, the nearest agreement was on eliminating "unnecessary regulation" as a contribution to price stability. President
Carter's economists were subsequently so inspired. The merits
of regulation may be debated. But the removal of all debatable
business regulation would not alter the consumer price index
by more than half a percent in half a century. And this most
economists know.

The third evasion is to promise to restore competition to the
economy — in effect to re-establish the structure of the econ-

omy in which the established fiscal and monetary remedies can work. Liberal economists, when all else fails, call for enforcement of the antitrust laws as an inflation remedy. It is the last wavering gasp of the bankrupt mind.

Bringing representatives of capital and labor together for agreement on price and wage policies in a smoke-filled Washington room is a more recent design for simulating action. As an isolated, unstructured effort, unbacked by serious government purpose, such consultations exploit only the constitutional right to free assembly.

The most elaborate evasion involves bringing moral pressure to bear on trade unions and corporations to moderate their wage claims and price increases. As this goes to press, this evasion is the one being favored. Under President Ford the Council on Wage and Price Stability was established to simulate action. It was denied all power except that of free speech. Under the ensuing Democratic administration the illusion of action was extended by appealing first to the political competence of Mr. Robert Strauss and then to the administrative enthusiasm of Dr. Alfred E. Kahn. Limits were set on permissible wage and price increases with no penalties against those that were deemed impermissible. That would have involved controversy. It was, perhaps, a mark of the improving public perception of such efforts that, except by those immediately involved, nothing was expected of them. Those involved were assumed to be serving not for the results but, more legitimately perhaps, for the pay.

Finally, there is a continuing use of prediction as a substitute for action. It is announced each month, as the inflation and unemployment figures are released, that everything will be better in the third quarter hence. The press does remain very tolerant of this evasion and duly reports it as though true.

One sees the glum position in which economists of comfortable inclination now find themselves. The stark, blunt fact is that orthodox monetary and fiscal policy give us not a choice between an unacceptable rate of inflation and an unacceptable level of unemployment but an unacceptable combination of the two.[2] Only the uniquely brave or reckless, whether liberal or conservative, can say that the combination of inflation and unemployment is something the system can suffer and survive. All must promise improvement. And, it will be seen, the techniques of evasion, though still much practiced, are running out. There remains only the ability of all in our profession, when in office, to believe that because the fates have been so wise as to place them there, they will, even in the absence of effort, rescue and provide. Conservative economists, if sufficiently archaic, have some justification for this theology; the full employment equilibrium is basic to the conservative creed, and inaction is the way to realize that equilibrium. Liberal economists must believe that they are of the chosen — that, as I've said often, God is a Keynesian Democrat.

All modern industrial countries are subject to the same tyrannical circumstance. In all, the large corporations, unions and numerous individuals have escaped the discipline of the market and gained power over their incomes. The exercise of this power drives up prices. When orthodox monetary and fiscal policy are used to arrest this upward thrust, it is production, not price, that is curbed. And with the curtailment of production goes curtailment of employment. Until unemployment is very great, perhaps up in the recession range of 10 percent or more of the labor force, unemployment and inflation coexist. Germany, Switzerland and Austria have gained substantial control over incomes through nationwide collective bargaining that limits increases to what can be afforded from stable prices. Britain has been working to the same end. The United States has still to master this disagreeable task. It's an ungrateful world.

2 It is, perhaps, proof either of progress or of the detachment to which the academic community can rise that the Ford Foundation has announced a series of grants for the study of this problem. (Ford Foundation Letter, April 1, 1977.) One of the studies is expected to last for five years.

Just as economists come into public office in unprecedented numbers, we discover that the economist's life was not meant to be a happy one. Anciently it has been said that ours is a subject that deals with choice. Now we discover that this is true even of public careers. They can be peaceful, fraudulent and fear-ridden. Or they can be innovative, successful and very controversial.

Certainly no President should be in doubt. If his economists are winning applause, inspiring confidence, avoiding acrimony, he should be deeply alarmed. The effects will spill over on him with politically fatal results. I am not reaching for paradox or exaggeration; the historical affirmation is complete. Three Presidents in the last fifty years have enjoyed supreme business confidence — Herbert Hoover, Richard Nixon and Gerald Ford. Mr. Nixon didn't finish his term. Mr. Hoover and Mr. Ford were the only two Presidents in this century who failed in their bids for a second term.

But more is at stake than presidential careers. Unless economists understand that our subject is intrinsically contentious — that what is good for the poorest of our people is best for economic performance but worst for gaining applause — economic policy will be a failure. We will have more tax cuts with heavy incidence on the always articulate affluent, tax cuts that bypass the terrible needs of our cities; that lodge themselves heavily into savings; that, when spent, affect most those markets where labor is already tight; and that thus make the greatest possible contribution to inflation. Unless the same economists proceed with the greatly contentious task of working out a system of income and price restraint over those who have gained some control of their prices and incomes — a task that must combine patient consultation with use of the powers of the state — inflation will either be uncontrolled or, as now, partially controlled by unemployment. And people, especially the poor, will come to wonder if having economists in office instead of lawyers is all that good.

3

The Higher Economic Purpose
of Women

IN THE NINETEEN-FIFTIES, for reasons that were never revealed to me, for my relations with academic administrators have often been somewhat painful, I was made a trustee of Radcliffe College. It was not a highly demanding position. Then, as now, the college had no faculty of its own, no curriculum of its own and, apart from the dormitories, a gymnasium and a library, no academic plant of its own. We were a committee for raising money for scholarships and a new graduate center. The meetings or nonmeetings of the trustees did, however, encourage a certain amount of reflection on the higher education of women, there being no appreciable distraction. This reflection was encouraged by the mood of the time at Harvard. As conversation and numerous formal and informal surveys reliably revealed, all but a small minority of the women students felt that they were a failure unless they were firmly set for marriage by the time they got their degree. I soon learned that my fellow trustees of both sexes thought this highly meritorious. Often at our meetings there was impressively solemn mention of our responsibility, which was to help women prepare themselves for their life's work. Their life's work, it was held, was care of home, husband and children. In inspired moments one or another of my colleagues would ask, "Is there anything else so important?"

Once, and rather mildly, for it was more to relieve tedium than to express conviction, I asked if the education we provided wasn't rather expensive and possibly also ill-adapted for these tasks, even assuming that they were combined with ultimate service to the New Rochelle Library and the League of Women Voters. The response was so chilly that I subsided. I've never minded being in a minority, but I dislike being thought eccentric.

It was, indeed, mentioned that a woman should be prepared for what was called a *second* career. After her children were raised and educated, she should be able to essay a re-entry into intellectual life — become a teacher, writer, researcher or some such. All agreed that this was a worthy, even imaginative design which did not conflict with *basic* responsibilities. I remember contemplating but censoring the suggestion that this fitted in well with the common desire of husbands at about this stage in life to take on new, younger and sexually more inspiring wives.

In those years I was working on the book that eventually became *The Affluent Society*. The task was a constant reminder that much information solemnly advanced as social wisdom is, in fact, in the service of economic convenience — the convenience of some influential economic interest. I concluded that this was so of the education of women and resolved that I would one day explore the matter more fully. This I have been doing in these last few years, and I've decided that while the rhetorical commitment of women to home and husband as a career has been weakened in the interim, the economic ideas by which they are kept persuaded to serve economic interests are still almost completely intact. Indeed, these ideas are so generally assumed that they are very little discussed.

Women are kept in the service of economic interests by ideas that they do not examine and that even women who are professionally involved as economists continue to propagate, often with some professional pride. The husband, home and family that were celebrated in those ghastly Radcliffe meetings are no

longer part of the litany. But the effect of our economic education is still the same.

Understanding of this begins with a look at the decisive but
little-perceived role of women in modern economic development and at the economic instruction by which this perception
is further dulled.

The decisive economic contribution of women in the developed industrial society is rather simple — or at least it so becomes once the disguising myth is dissolved. It is, overwhelmingly, to make possible a continuing and more or less unlimited
increase in the sale and use of consumer goods.

The test of success in modern economic society, as all know,
is the annual rate of increase in Gross National Product. At
least until recent times this test was unquestioned; a successful
society was one with a large annual increase in output, and
the most successful society was the one with the largest increase.
Even when the social validity of this measure is challenged, as
on occasion it now is, those who do so are only thought to be
raising an interesting question. They are not imagined to be
practical.

Increasing production, in turn, strongly reflects the needs of
the dominant economic interest, which in modern economic
society, as few will doubt, is the large corporation. The large
corporation seeks relentlessly to get larger. The power, prestige,
pay, promotions and perquisites of those who command or who
participate in the leadership of the great corporation are all
strongly served by its expansion. That expansion, if it is to be
general, requires an expanding or growing economy. As the
corporation became a polar influence in modern economic life,
economic growth became the accepted test of social performance. This was not an accident. It was the predictable acceptance of the dominant economic value system.

Economic growth requires manpower, capital and materials
for increased production. It also, no less obviously, requires

increased consumption, and if population is relatively stable, as in our case, this must be increased per-capita consumption. But there is a further and equally unimpeachable truth which, in economics at least, has been celebrated scarcely at all: just as the production of goods and services requires management or administration, so does their consumption. The one is no less essential than the other. Management is required for providing automobiles, houses, clothing, food, alcohol and recreation. And management is no less required for their possession and use.

The higher the standard of living, that is to say the greater the consumption, the more demanding is this management. The larger the house, the more numerous the automobiles, the more elaborate the attire, the more competitive and costly the social rites involving food and intoxicants, the more complex the resulting administration.

In earlier times this administration was the function of a menial servant class. To its great credit, industrialization everywhere liquidates this class. People never remain in appreciable numbers in personal service if they have alternative employment. Industry supplies this employment, so the servant class, the erstwhile managers of consumption, disappears. If consumption is to continue and expand, it is an absolute imperative that a substitute administrative force be found. This, in modern industrial societies, is the function that wives perform. The higher the family income and the greater the complexity of the consumption, the more nearly indispensable this role. Within broad limits the richer the family, the more indispensably menial must be the role of the wife.

It is, to repeat, a vital function for economic success as it is now measured. Were women not available for managing consumption, an upper limit would be set thereon by the administrative task involved. At some point it would become too time-consuming, too burdensome. We accept, without thought, that a bachelor of either sex will lead a comparatively simple existence. (We refer to it as the bachelor life.) That is because the administrative burden of a higher level of consumption, since it must be assumed by the individual who consumes, is a limiting factor. When a husband's income passes a certain level, it is expected that his wife will be needed "to look after the house" or simply "to manage things." So, if she has been employed, she quits her job. The consumption of the couple has reached the point where it requires full-time attention.

Although without women appropriately conditioned to the task there would be an effective ceiling on consumption and thus on production and economic expansion, this would not apply uniformly. The ceiling would be especially serious for high-value products for the most affluent consumers. The latter, reflecting their larger share of total income — the upper 20 percent of income recipients received just under 42 percent of all income in 1977 — account for a disproportionate share of total purchases of goods. So women are particularly important for lifting the ceiling on this kind of consumption. And, by a curious quirk, their doing so opens the way for a whole new range of consumer products — washing machines, dryers, dishwashers, vacuum cleaners, automatic furnaces, sophisticated detergents, cleaning compounds, tranquilizers, pain-relievers — designed to ease the previously created task of managing a high level of consumption.

Popular sociology and much associated fiction depict the extent and complexity of the administrative tasks of the modern diversely responsible, high-bracket, suburban woman. But it seems likely that her managerial effectiveness, derived from her superior education, her accumulating experience as well as her expanding array of facilitating gadgetry and services, keeps her more or less abreast of her increasingly large and complex task. Thus the danger of a ceiling on consumption, and therefore on economic expansion, caused by the exhaustion of her administrative capacities does not seem imminent. One sees here, more than incidentally, the economic rationale, even if it was unsuspected for a long time by those involved, of the need for a superior education for the upper-bracket housewife. Radcliffe prepared wives for the higher-income family. The instinct that this required superior intelligence and training was economically sound.

The family of higher income, in turn, sets the consumption patterns to which others aspire. That such families be supplied with intelligent, well-educated women of exceptional managerial competence is thus of further importance. It allows not only for the continued high-level consumption of these families, but it is important for its demonstration effect for families of lesser income.

That many women are coming to sense that they are instruments of the economic system is not in doubt. But their feeling finds no support in economic writing and teaching. On the contrary, it is concealed, and on the whole with great success, by modern neoclassical economics — the everyday economics of the textbook and classroom. This concealment is neither conspiratorial nor deliberate. It reflects the natural and very strong instinct of economics for what is convenient to influential economic interest — for what I have called the convenient social virtue. It is sufficiently successful that it allows many hundreds of thousands of women to study economics each year without their developing any serious suspicion as to how they will be used.

The general design for concealment has four major elements:

First, there is the orthodox identification of an increasing consumption of goods and services with increasing happiness. The greater the consumption, the greater the happiness. This proposition is not defended; it is again assumed that only the philosophically minded will cavil. They are allowed their dissent, but, it is held, no one should take it seriously.

Second, the tasks associated with the consumption of goods are, for all practical purposes, ignored. Consumption being a source of happiness, one cannot get involved with the problems in managing happiness. The consumer must exercise choice; happiness is maximized when the enjoyment from an increment of expenditure for one object of consumption equals that from the same expenditure for any other object or service. As all who have ever been exposed, however inadequately, to economic instruction must remember, satisfactions are maximized when they are equalized at the margin.

Such calculation does require some knowledge of the quality and technical performance of goods as well as thought in general. From it comes the subdivision of economics called consumer economics; this is a moderately reputable field that, not surprisingly, is thought especially appropriate for women. But this decision-making is not a burdensome matter. And once the decision between objects of expenditure is made, the interest of economics is at an end. No attention whatever is given to the effort involved in the care and management of the resulting goods.[1]

The third requisite for the concealment of women's economic role is the avoidance of any accounting for the value of household work. This greatly helps it to avoid notice. To include in the Gross National Product the labor of housewives in managing consumption, where it would be a very large item which would increase as consumption increases, would be to invite thought on the nature of the service so measured. And some women would wonder if the service was one they wished

[1] There is a branch of learning — home economics or home science — that does concern itself with such matters. This field is a nearly exclusive preserve of women. It has never been accorded any serious recognition by economists or scholars generally; like physical education or poultry science, it is part of an academic underworld. And home economists or home scientists, in their natural professional enthusiasm for their subject matter and their natural resentment of their poor academic status, have sought to elevate their subject, homemaking, into a thing of unique dignity, profound spiritual reward, infinite social value as well as great nutritional significance. Rarely have they asked whether it cons women into a role that is exceedingly important for economic interest and also highly convenient for the men and institutions they are trained to serve. Some of the best home economists were once students of mine. I thought them superbly competent in their commitment to furthering a housewifely role for women.

to render. To keep these matters out of the realm of statistics is also to help keep them innocuously within the sacred domain of the family and the soul. It helps sustain the pretense that, since they are associated with consumption, the toil involved is one of its joys.

The fourth and final element in the concealment is more complex and concerns the concept of the household. The intellectual obscurantism that is here involved is accepted by all economists, mostly without thought. It would, however, be defended by very few.

The avowed focus of economics is the individual. It is the individual who distributes her or his expenditures so as to maximize satisfactions. From this distribution comes the instruction to the market and ultimately to the producing firm that makes the individual the paramount power in economic society. (There are grave difficulties with this design, including the way in which it reduces General Motors to the role of a mere puppet of market forces, but these anomalies are not part of the present story.)

Were this preoccupation with the individual pursued to the limit, namely to the individual, there would be grave danger that the role of women would attract attention. There would have to be inquiry as to whether, within the family, it is the husband's enjoyments that are equalized and thus maximized at the margin. Or, in his gallant way, does he defer to the preference system of his wife? Or does marriage unite only men and women whose preference schedules are identical? Or does marriage make them identical?

Investigation would turn up a yet more troublesome thing. It would be seen that, in the usual case, the place and style of living accord with the preferences and needs of the member of the family who makes the money — in short, the husband. Thus, at least partly, his titles: "head of the household," "head of the family." And he would be seen to have a substantial role in decisions on the individual objects of expenditure. But the management of the resulting house, automobile, yard, shopping and social life would be by the wife. It would be seen that this arrangement gives the major decisions concerning consumption extensively to one person and the toil associated with that consumption to another. There would be further question as to whether consumption decisions reflect with any precision or fairness the preferences of the person who has the resulting toil. Would the style of life and consumption be the same if the administration involved were equally shared?

None of these questions gets asked, for at precisely the point they obtrude, the accepted economics abruptly sheds its preoccupation with the individual. The separate identities of men and women are merged into the concept of the household. The inner conflicts and compromises of the household are not explored; by nearly universal consent, they are not the province of economics. The household, by a distinctly heroic simplification, is assumed to be the same as an individual. It thinks, acts and arranges its expenditures as would an individual; it is so treated for all purposes of economic analysis.

That, within the household, the administration of consumption requires major and often tedious effort, that decisions on consumption are heavily influenced by the member of the household least committed to such tasks, that these arrangements are extremely important if consumption is to expand, are all things that are thus kept out of academic view. Those who study and those who teach are insulated from such adverse thoughts. The concept of the household is an outrageous assault on personality. People are not people; they are parts of a composite or collective that is deemed somehow to reflect the different or conflicting preferences of those who make it up. This is both analytically and ethically indefensible. But for concealing the economic function of women even from women it works.

One notices, at this point, an interesting convergence of economics with politics. It has long been recognized that women are kept on political leash primarily by urging their higher commitment to the family. Their economic role is also concealed and protected by submerging them in the family or household. There is much, no doubt, to be said for the institution of the family. And it is not surprising that conservatives say so much.

In modern society power rests extensively on persuasion. Such reverse incentives as flogging, though there are law-and-order circles that seek their revival, are in limbo. So, with increasing affluence, is the threat of starvation. And even affirmative pecuniary reward is impaired. For some, at least, enough is enough — the hope for more ceases to drive. In consequence, those who have need for a particular behavior in others resort to persuasion — to instilling the belief that the action they need is reputable, moral, virtuous, socially beneficent or otherwise good. It follows that what women are persuaded to believe about their social role and, more important, what they are taught to overlook are of prime importance in winning the requisite behavior. They must believe that consumption is happiness and that, however onerous its associated toil, it all adds up to greater happiness for themselves and their families.

If women were to see and understand how they are used, the consequence might be a considerable change in the pattern of their lives, especially in those income brackets where the volume of consumption is large. Thus, suburban life sustains an especially large consumption of goods, and, in consequence, is especially demanding in the administration required. The claims of roofs, furniture, plumbing, crabgrass, vehicles, recreational equipment and juvenile management are all very great. This explains why unmarried people, regardless of income, favor urban living over the suburbs. If women understood that they are the facilitating instrument of this consumption and were led to reject its administration as a career, there would, one judges, be a general return to a less demanding urban life.

More certainly there would be a marked change in the character of social life. Since they are being used to administer consumption, women are naturally encouraged to do it well. In consequence, much social activity is, in primary substance, a competitive display of managerial excellence. The cocktail party or dinner party is, essentially, a fair, more refined and complex than those at which embroidery or livestock are entered in competition but for the same ultimate purpose of displaying and improving the craftsmanship or breed. The cleanliness of the house, the excellence of the garden, the taste, quality and condition of the furnishings and the taste, quality and imagination of the food and intoxicants and the deftness of their service are put on display before the critical eye of those invited to appraise them. Comparisons are made with other exhibitors. Ribbons are not awarded, but the competent administrator is duly proclaimed a good housekeeper, a gracious hostess, a clever manager or, more simply, a really good wife. These competitive social rites and the accompanying titles encourage and confirm women in their role as administrators and thus facilitators of the high levels of consumption on which the high-production economy rests. It would add measurably to economic understanding were they so recognized. But perhaps for some it would detract from their appeal.

However, the more immediate reward to women from an understanding of their economic role is in liberalizing the opportunity for choice. What is now seen as a moral compulsion — the diligent and informed administration of the family consumption — emerges as a service to economic interests. When so seen, the moral compulsion disappears. Once women see that they serve purposes which are *not* their own, they will see that they can serve purposes which *are* their own.

4

The Conservative Majority Syndrome

In 1978, following the adoption of the Jarvis-Gann tax limita-
tion amendment (Proposition 13) in California, a giant wave
of conservative enthusiasm and evangelism was held to be
sweeping the country. This article was written two years earlier,
before the 1976 election. A conservative mood was *then* held
to be enveloping the Republic.

MY THOUGHT in this essay is to explain in a scientific way the
powerful and wonderfully persistent forces that, recurrently,
seek to persuade us that conservatism is the wave of the American future. These instruments of persuasion are brought to
bear at all times and with much success on the Congress. And
before all elections they are directed with great energy at the
Democratic Party and the nation at large.

The purpose does not vary. It is to persuade all susceptible
citizens, but legislators and candidates in particular, that the
country has, at long last, moved sharply to the right. Politically
speaking, there are no poor, no aged, no sick, no black, no other
minorities, no people seriously squeezed by inflation, not many
for whom unemployment is a major issue, no one whatever
whose health, education, food, shelter, protection from economic abuse or exploitation, or even whose survival itself depends on the services of government. Instead there are in this
Republic only indignant taxpayers deeply angry about the will-

ful idleness of public servants and the unemployed. The only sophisticated policy is their appeasement. The prime enemy of the people is the government, save as it involves itself in the exigent and increasing needs of national defense and those of bankrupt but still meritorious corporations. The ultimate tendency (and hope) of our politics is toward two equally conservative parties competing for the votes of the great conservative electorate.

We should have a name for this phenomenon; I propose that it be called "The Conservative Majority Syndrome." In an earlier time it would have been called "The Dayton Housewife Discovery." That excellent woman, the invention of Messrs. Scammon and Wattenberg in the political campaign of 1972, was also unblack, unpoor, definitely uninterested in anything as unrefined as women's rights and, you can be sure, deeply concerned about taxes. She was the typical American. Americans and Dayton were greatly libeled.

The success of the conservative majority syndrome depends on four motivating factors, all powerful in our time. The first is the susceptibility of much of our political comment and many of our political commentators and columnists, most notably the great men of television, to the rediscovery of the wheel. Its special manifestation is the recurrent analysis, offered each time as a breathtaking revelation of only slightly less than scriptural impact, that people of means would rather not pay taxes. This is never combined, most regrettably, with the companion revelation that people of means are infinitely more articulate than anyone else. So it is not noticed that, by its sheer volume, the voice of relative affluence in our land gets mistaken for the voice of the masses. Anyone in doubt on this point should try to recall how many welfare recipients were heard on the question of the profligacy of New York City in its troubled days as compared with the volume of expression that emanated from the Chase Manhattan Bank, the investment banking house of

Lazard Freres and then Secretary of the Treasury William Simon. The Westchester County budget, it was announced in the autumn of 1975, would rise by 22 percent in 1976. In those mostly pleasant and affluent precincts this was held to be the result of changing population structure, inflation and recession. In New York City the same increase would have been attributed entirely to the people on welfare and to the unions.

The second support to the conservative majority syndrome lies in the deep desire of politicians, Democrats in particular, for respectability — their need to show that they are individuals of sound, confidence-inspiring judgment. And what is the test of respectability? It is, broadly, whether speech and action are consistent with the comfort and well-being of people of property and position. A radical is anyone who causes discomfort or otherwise offends such interests. Thus, in our politics, we test even liberals by their conservatism. Nothing, in fact, should be so damaging to a liberal as an approving editorial in the *Wall Street Journal*. Were it only so!

There is a self-reinforcing aspect to this particular conservative thrust that works with special effect in Washington. There some conservative legislator to whom the economics of Andrew Mellon would have seemed advanced examines a probable action and concludes that it will cause pain to the privileged. This sorrow he then identifies with popular outrage, and he relays his views to either Mr. Evans or Mr. Novak. These scholars tell of the intended action and cite as general the belief that it will be a source of much indignation, if not of mass anguish. The progenitor reads their story and is affirmed in his fears. So are others. It is, I once concluded, the only completely successful closed-circuit system for recycling garbage that has yet been devised.

The third strength of the conservative majority syndrome lies in the superb tactical position of conservatives when they are in power. They can attack the government for indifference, callousness or incompetence and then justify the attack by making it so. This was uniquely the achievement of Richard Nixon. Government, we might recall, was not thought callous, indifferent or incompetent in the days of Roosevelt or Truman or of Kennedy or Lyndon Johnson.

We must never minimize the importance of good public management and administration, a dull but important business. It has not had the attention from liberals that it deserves. But let us not join in the currently fashionable tendency to defame either government or those who work for government. The civil service of the United States is as honest, effective and certainly as innovative as that of any other country. It was stubbornly honest people in the Department of Justice, the FBI, the IRS and elsewhere who frustrated the tax evasion, obstruction of justice, subornation of perjury and constitutional subversion of Richard Nixon and the unimaginative felonies of Spiro Agnew. Had our public servants not been honest, our case would have been hopeless.

The final buttress to the conservative majority syndrome is, oddly, many of the economists of the liberal establishment. Here I must proceed tactfully; I am speaking of conscientious people, excellent citizens, good parents, kind friends. But time, alas, in its ineluctable march, has made them pillars of the conservative edifice. This is partly because some of them yearn, as always, for Establishment applause and test their performance by its volume.

Also many of these scholars serve within an institutional framework; and some are sustained visibly and directly by the banks and corporations they assist, advise or, on occasion, operate. Out of institutional identification comes a largely innocent and natural desire to avoid embarrassing one's colleagues and employers by urging policies that are at odds with the respectable view. And out of corporate service comes the less innocent desire to protect one's income. Much of what was once called liberal economics has become very conservative in our time.

One manifestation of this conservatism in contemporary economics is the unwillingness to proceed with the painful actions that are required to control inflation and reduce unemployment.[1] Better, especially when in power, to sit quietly and hope. Or, when pressed, say that the real need is for more expansion, jobs and economic growth for which the required policies — easier money and tax reduction — are pleasant to prescribe. So far as this guidance is accepted, our political life is then divided between conservatives who prefer unemployment to inflation but do not say so and conservatives who prefer inflation to unemployment but do not say so.

There is no future for liberals in such a debate. Unemployment hurts a smaller number of people a lot; inflation hurts a larger number of people a little. We cannot defend either. It is not possible to persuade people — as some liberals would — that inflation is an overrated evil. For the person wondering how to stretch limited income or savings over urgent needs, inflation is real. And people's dislikes are as they are. It is not the liberals' business to try to say they are wrong.

A further support to conservatism in established economics — and this is a very important matter — is in its favored formula for expanding the economy. In recent years its remedy has always been the reduction of federal income taxes. This prescription is offered all but automatically whenever the economy is operating at less than full capacity, which, of course, has been frequently. Nothing has so effectively played into the hands of conservatives.

There was always the danger that tax reduction would be

[1] See the earlier essay, "Economists and the Economics of Professional Contentment."

part of a disguised effort to limit public expenditures — excluding, of course, defense and those in which the modern large corporation has a prime interest. Gerald Ford, to his credit, removed the disguise. He asked for tax reduction and made it conditional on expenditure reduction. Those who had feared this result should be grateful to him for clarifying a difficult point.

We should be grateful as well to Professor Milton Friedman, a man of inconvenient honesty, who also made the conservative position wonderfully clear. He has reported himself in favor of a federal tax cut at all times as the only way to exert effective pressure on Congress to keep down federal spending.

Public expenditures are progressive in their incidence — they have their greatest effect on those who are least able to provide income, schools, health services, housing or police protection for themselves. So tax reductions that pave the way for expenditure reductions — outlays for defense services and corporate spending always being exempted — have a special impact on the poor.

It is federal income taxes that are reduced. These, in substantial though not exclusive part, are on middle- and upper-income brackets of the personal income tax and on corporations. This reduction in recent years has been at a time when states and cities, in consequence of the recession a tax reduction was to cure, were being forced to raise *their* taxes. Local taxes, invariably, are more regressive. In 1975, while the House Ways and Means Committee was discussing the extension of a cut in federal income taxes, Secretary of the Treasury Simon, in a typically compassionate mood, proposed that New York City raise its sales tax to 10 percent. State and city taxes were raised instead. And the services of a city already deep in public squalor were curtailed.

In the same week that the Ways and Means Committee reported out that tax cut bill, the Commonwealth of Massachusetts agreed on a tax increase of $364 million, all of it in relatively regressive taxes, for no others were available. (Two hundred and eighteen million dollars would come from taxes on sales and meals.) The *Boston Globe,* noting the juxtaposition of these events, said it made no sense. They were right. There was, needless to say, no economic stimulation in a shift from progressive federal to regressive state and local taxes. The only joy was for the affluent. Such is the service of the established economics to conservatism.

5

The Multinational Corporation: How to Put Your Worst Foot Forward or in Your Mouth

IN THE LAST THIRTY YEARS no institution has so intruded itself on the economic landscape as the multinational corporation. None has provoked so much discussion or been the subject of such obsessive concern,[1] and almost every reference has had in it a note of anxiety or even hostility.

Such hostility, the existence of which no corporate executive will deny, rests, most executives will plead, on public misunderstanding. They look at themselves, their colleagues, their hours of work, their service to customers and tax collectors, and at their families, churches, charities and largely innocent recreations, and ask if they are really wicked men. Their answer, quite rightly in the main, is no. But the executive of the multinational corporation should be in no doubt as to who is responsible for the misunderstanding. He is misunderstood because, usually after some deliberation and often with some passion, he insists on misunderstanding himself. He puts forward an

[1] The literature of the multinational enterprise must have the largest proportion of spoil to ore of any field of economic writing, theoretical and mathematical model-building possibly apart. I've cited it sparsely in this essay, but I would like to make special mention of the recent book by my colleague Raymond Vernon, *Storm over the Multinationals: The Real Issues* (Cambridge: Harvard University Press, 1977), which provides admirable background for the matters here examined.

explanation of himself and his company that is based on outworn economic theory with a strong aspect of theology. It is, in consequence, almost wholly implausible. No man, to paraphrase Keynes, is so much a slave of such defunct economists.

There are difficulties with a defense of the multinational corporation that is based on reality. They arise because such a defense must concede what for so long was so indignantly denied. Nonetheless the time has come when a realistic approach is not only wiser but even, I would judge, unavoidable.

I've lived most of my life in close professional association with the large modern corporation. I have come to accept that it is inevitable. Among other things, it combines energies, experience, engineering and scientific and other specialties for results far beyond the abilities or, on occasion perhaps, even the conceptual reach of any individual. The deeply collectivist character of this effort is in the sharpest contrast with the individualist economic, social and political case that it makes for itself. I would like in this essay to look first at the circumstances that explain the rise of the multinational corporation; then at the exceptionally primitive ideas that now guide its defense; then at the case as it might be made; and, finally, at the management behavior that this defense requires.

The multinational corporation — the company that extends its business operations under one guiding direction to two or more countries — gained major public attention in the thirty years following World War II. During these years it moved heavily into industrial operations — into the manufacturing or processing of goods, their marketing and sales. In its multinationality it had, however, three antecedents in non-manufacturing fields, all of them exceedingly ancient. These had something, in fact much, to do with the reputation of, and the reaction to, the industrial corporation when it became the dominant corporate form operating across national frontiers.

First: Since the rise of the modern national state, banking

operations have been carried on across national borders, partly to finance international trade, more significantly for present purposes to take capital from countries of relative abundance to those of relative scarcity and consequent higher return. Second: The resource industries, mining in particular, have long been multinational. These industries have gone, as a matter of course, to develop operations where the ore, natural products or, latterly, the oil were available. Third: Multinational trading corporations have traditionally brought the products of the industrial countries to the economically more backward lands.

In each of these antecedent forms there was a major element of adverse reputation that became part of the legacy of the modern multinational. The international banker had a notably sinister reputation: his concern for his capital transcended national interest and caused him on occasion to consort with the enemies of his sovereign. He was invariably richer than the borrower; and he was thought to exploit the profligacy or unwisdom of the latter or, in any case, was celebrated for so doing. In the absence of other available ill fame, anti-Semitism could often be evoked.

Resource industries, inevitably, were held to be robbing the countries where they operated of their patrimony — their natural wealth. They were thought to exploit local labor, though partly because it had an even lower local opportunity cost. And they did bend weak local governments to their purposes.

In India, Indonesia and Indochina the trading corporations were intimately linked with colonialism and in China with foreign penetration and domination. Socialists still celebrate them as the *compradors*.

Thus, like so many children, the multinational industrial corporation was unwise in the choice of its parents and is visited with their sins.

The case for the modern multinational begins with its considerable differences from its three antecedents, something that

has been but rarely noticed. The older multinational operations mostly traded between the rich countries and the poor.
They bridged the gap between those countries with more
capital than they needed and those with less, those with manufacturing industries and those countries that could supply
only raw materials and agricultural products, those with finished products to sell and those that, being devoid of modern
industry, could only buy.

These antecedent forms have themselves changed. International banking continues, but operations between the rich
countries and the poor are not what now evokes the modern
image of the multinational banking corporation. Most of its
operations — and to a singular degree its better recent fortune,
as the experience of the big New York banks affirms — have
been between the industrial countries. Raw materials, including oil, and some tropical food products still come from the
poor countries to the rich. But the greatest suppliers of wheat,
feed grains, coal, wood and wood pulp and cotton fiber are the
two North American countries, the United States and Canada.
(If to be part of the Third World is to be a hewer of wood and
a supplier of food and natural products, the United States and
Canada are, by a wide margin, the first of the Third World
countries and should vote accordingly in the United Nations.)
And the trading corporation, has, of course, receded greatly in
relative importance.

Thus even the older forms of multinational operations are
now mainly, though of course not exclusively, between the
developed countries. This, however, is most strongly the case
in the new and burgeoning area of multinational operations —
manufacturing. And this is for a very special and much neglected reason.. The multinational corporation is the nearly
inescapable accommodation to international trade in modern
capital and consumer goods. One must emphasize this central
point.

The products principally traded in past times — foodstuffs,

cotton, cotton textiles, wool, coal, steel rails — required no connection or communication between producer and consumer. They could be and were shipped and sold through intermediaries, exporters and importers; the producer never saw the user. Products were also sold in the market, and, very often, the market price and sometimes the destination were unknown when the goods were shipped. Before 1914, grain ships approaching Land's End from North America regularly received a signal as to where they should proceed in Europe for the best price. For this trade, and similarly for simple unbranded manufactures such as cotton greige goods, there was no role for a multinational enterprise, and we need scarcely remind ourselves that the original producers were, in the vast majority of cases, very small.

To a greater extent than one likes to think, this view of international trade still rules in the economics courses and textbooks. It is one of the several reasons why formal economics has had difficulty in coming to terms with the multinational firm.

The modern industrial enterprise, on the other hand, has products that must be marketed. Only in the textbooks is the consumer left to his or her sovereign decision. There also must be control over final prices; it is elementary that General Motors and Volkswagen do not ship their vehicles for sale, in the manner of wheat, for what the market will bring. Both the merchandising and the market regulation require a well-controlled sales organization, as do the instruction, repair and service that many modern products need and often receive.

The modern industrial enterprise has, as well, numerous needs from the government of the country in which its products are sold — far more things than the mindless litany of private enterprise admits. The government is an important customer. It is also the source of the airports, airways, highways, television channels, telephone communications, weapons orders and financial aid without which many, perhaps most, modern industrial products cannot be produced or sold. And it is the source, as well, of a wide and increasing range of permissions and restrictions governing the sale and use of these products.

So the modern manufacturing corporation must have an intimate relationship — dependent, symbiotic, sometimes suborning — with the modern government. It must, in short, have influence and power in its own markets — over its prices, costs and means of consumer persuasion. And similarly it is impelled to seek influence and power in the modern state.

Power is a word that, in deference to conventional economics and its more general public effect, every business executive seeks to avoid. There is something exceptionally improper about its possession. But power is indispensable to the operations of the modern large corporation. It must have power over its prices; its planning operations could not possibly survive the kind of price instability that characterizes most small entrepreneurial industry. It must have a measure of control over its earnings; these are a vital source of capital, as all accept. It has a long production period involving heavy investment; it must, accordingly, have power to persuade the consumer to want the product that eventually emerges. It lives with the consent of and by the support of the state; so living, it seeks to influence the decisions of the government.

This exercise of power is not a matter of choice but of necessity. It is true of the national corporation. And it is wholly and equally true of the large multinational corporation.

Along with the omnipresent impulse to growth, the need to exercise extra-market power is the prime impulse to multinationality. A corporation in New York or Pittsburgh that is confined in its operations to the United States cannot bring power to bear in France to protect its prices in French markets, develop its dealer organization, defend or advance its position with the French consumer and protect or advance its position with the French government. Yet if the corporation is to do business in France, it will have to bring such power to bear, unconventional economic liturgy and official corporate rhetoric notwithstanding. It must have a base or presence there for the exercise of the market power and the public power needed to develop and protect its market position and its public position. The extent of the influence and power that can be so exercised will be in very rough proportion to the scale of this national presence.

Such facts explain the corporate tendency to go on from international sales representatives to a sales and marketing organization to a more fully integrated international development. This is the dynamic of modern international trade. This dynamic is reinforced by several further factors.

Where advanced technology is involved, only multinational operations realize the full economies of scale — the returns on one development cost are realized in several or numerous national markets.[2] And multinational intrusion by the corporations of one country also forces reciprocal action by those intruded upon. If there is no such response, markets are lost with no chance of compensatory gain. Such riposte may be necessary as well to preserve what economists call the "oligopolistic equilibrium" — meaning that the intruding corporation will not cut prices, for, if it does so, it will be vulnerable to the same action in reciprocal form in its own home country.

The forces just adumbrated are not transitory or casual. They will continue, and so, accordingly, will the development of multinational enterprise. This development would invite less anxiety if those involved were to learn to defend the multinational corporation with some semblance of plausibility and if they were to avoid behavior that cannot be defended. Before they can do this, they should examine the present defense.

Not seeing that the multinational corporation is the necessary manifestation of international trade in manufactured prod-

2 As Peter J. Buckley and Mark Casson have shown in *The Future of the Multinational Enterprise* (London: Macmillan, 1976), there is a rough correlation between the research intensivity of an industry and the extent of multinational operations.

ucts, its spokesmen rarely make this point. Failing also to see that power is an essential aspect of corporate development, they urge with great implausibility that none is exercised. Instead they argue that the foreign corporation — a guest in the house — must be most cautious about influencing the habits, tastes, markets or public opinion of the host country, and especially so the actions of its government.

The denial that the corporation has power derives from its acolyte economists, the neoclassical scholars who provide the accepted economic and business theology. This holds that markets are made by numerous entrepreneurs, and all are ruled by the market. All accept and should accept the instruction of the sovereign consumer. Anything else is monopoly or oligopoly or monopolistic competition — a market imperfection — and thus an aberration. And as market power is absent, so also is public power. The classic corporation, like the citizen, can petition the state; any further exercise of power is irregular and improper. Its association is "with words like free trade and free enterprise and laissez-faire, [which] holds that business is politically neutral, existing only to satisfy the economic desires of the world's people."[3] To concede that a corporation has market or public power in the neoclassical economic tradition is to confess to impropriety or antisocial behavior.

Having taken the position that they have no power, the executives of the multinational enterprise then go on to exercise power in the most visible possible fashion. An increasingly attentive public observes, and it sees, therewith, the mendacious or ridiculous character of the corporate defense. Having denied that he exercises power, having conceded that his possession of such power is improper and having conceded its especial inappropriateness for a foreign-based enterprise, the executive of the multinational corporation proceeds with the

[3] "People, Politics and Productivity: The World Corporation in the 1980's." Address by Walter B. Wriston in London on September 15, 1976. Mr. Wriston, the head of Citicorp and Citibank, here makes another common and deeply damaging error. That is to assume that being the leader of a large corporation somehow compels belief for wholly implausible assertions.

utmost reliability to convict himself by his own actions. The day after he disavows market power, explains that his corporation is wholly subordinate to the market and denies any thought of interfering with the government, his company announces a price increase, launches a major advertising campaign to alter consumer preference and is revealed to have brought public pressure in some world capital in favor of or against some regulation, or on behalf of some new weapon, or to have contributed substantially to some domestic or foreign politician not as an act of philanthropy.

The only reasonable defense of the multinational corporation is now the truth. That the corporation has power must be conceded. The only durable defense then is to hold that such exercise of power is inevitable and, if subject to proper guidance and restraint, can be socially useful.

A related line of defense of the multinational has been to seek to conceal or deny its foreign base and origin. The literature of multinational operations stresses this point *ad nauseam;* so, one judges, do the internal manuals, handbooks and lectures. The corporation operating in a foreign country should always keep what executives with a less than original gift for language unite in calling "a low profile." It should also make maximum use of local technical and managerial personnel; the number of people from the country of origin should be kept to a minimum. The corporation must always be a good local citizen.

This defense is also implausible, and self-defeating as well. No one worth persuading will ever believe that General Motors in its origins is anything but American, BP anything but British, Fiat other than Italian. Persuasion is possible but not against elementary common sense.

And there is also question as to why, in an adult industrial world, there should be apology for this kind of international development. International trade had always to be defended against those who saw only its costs, never its advantages; who saw only the intrusion of foreign competitors, never the resulting efficiency in supply or products or the reciprocal gains from greater exports. The multinational corporation comes into existence when international trade consists of modern technical, specialized or uniquely styled manufactured products. Accordingly, it should be defended, as international trade was defended, for its contribution to efficiency in production and marketing, to living standards and to reciprocal opportunities in other lands for the enterprises of the host country. This affirmative defense is excluded if a negative defense concedes, in effect, that there is something wrong with being a foreign corporation. And this is precisely what talk about "a low profile" or minimizing the foreign presence concedes. It would be a poor defense of foreign trade to pretend that French wines in the United States or American cotton in Britain were really domestic products or somehow not foreign. This tactic would concede that there was something wrong with international trade in these commodities. The "low profile" strategy of the multinational corporation is the equivalent.

A third defense of the multinational corporation, more often implicit than explicit, has been that it reflects special aptitude in the development of management skills. The multinational reflects, merely, the extension of this supposedly American skill to the other industrial countries of the world.

This defense lurks strongly in the consciousness of many American executives, but it also owes something to the flattery, genuine or contrived, of their foreign colleagues in the business world. In the prose of one British leader, the "track record of the United States — in business as well as government — makes legitimate your leadership of the rest of us in tackling the world's economic problems."[4] The multinational corporation,

---

[4] Sir Reay Geddes (Chairman, Dunlop Holdings, Ltd.), "A Multinationalist's View," *The Future Role of Business in Society* (New York: The Conference Board, 1977), p. 17.

so viewed, is the natural result of American economic leadership, the natural expression of a basic American skill, the natural flowering of the American economic system.

The recent history of multinational development solidly refutes this case. It is possible that American companies give more prestige to business managers than do other countries, Japan apart. No other country is so deferentially frightened of anything that is thought to be damaging to business confidence. We can certainly take credit for pioneering in the field of business education and the theory and practice of management. But it is nonsense to suppose that there is an American business or managerial genius, that there is anything known to anyone from Illinois or California that cannot equally be learned by anyone from England, Switzerland or Sweden. In some aspects of multinational operations, in fact, there is an American inferiority complex based, as with the Japanese, on a congenital inadequacy in languages.

However rewarding to national or corporate vanity, the notion that the multinational corporation reflects American achievement or leadership is uniquely poisonous. Those who take credit for it as an American achievement must then defend it against the counterpart charge that it is an instrument of American economic, cultural or military imperialism.

Let me now turn to the affirmative arguments — the case that, with candor and a modest exercise of practical intelligence, could be made.

This requires, first of all, that the power of the multinational corporation, like that of the national corporation, be conceded. The point to be stressed is that this power is deployed, on balance, for socially useful ends; where it is not, it must expect to be restrained by the state.

The social and practical achievements of the international corporation should then be mentioned. Thus, governments spend notable energy seeking to improve the performance of national industries — agriculture, housing, health services. Success is far from complete. The supply of automobiles, chemicals, computers, television sets, tobacco, alcohol and other products of multinational origin is rarely a source of complaint as to quantity or cost. Other achievements of the multinational corporation are the protection of international trade, the reduction of international friction, the creation of an international civil service, the fostering of local managerial autonomy and the more rapid dissemination of technology around the world.

Business firms, large and small, have never been indifferent to government policy on international trade. They have not hesitated to use their power to influence such policy. It has rarely been noticed or urged that, as the multinational manufacturing corporation has grown in influence, the importance of tariff barriers has receded. The barriers are not needed by the multinational firm; for integrating and rationalizing operations as between plants in different countries, they are a nuisance. Foreign competition also diminishes greatly in terror when you own the relevant competitor. So power that was once deployed on behalf of tariffs is no longer exercised. On balance, it is deployed against them. The European Economic Community did not come into existence because of a sudden access of economic enlightenment after World War II; it came into existence because modern corporate and multinational organizations had made the old boundaries and barriers obsolete. European farmers, with their different, earlier and essentially classical economic structure, would never have created the Common Market. They are now the source of most of its disputes and nearly all of its crises.

The great reduction of the importance of tariffs, except in agriculture and a few other national industries, has, in turn, removed or reduced an important source of international friction. This reduction is one substantial accomplishment of the multinational corporation, and it has made a yet more profound contribution to international amity. The absence of economic conflict, like peace in general, is so unobtrusive that we often do not notice it. It is almost never a part of the multi-nationals' defense.

In the last century national industries, notably in steel, coal, ship- and machine-building, were the natural allies of national governments in the development of armaments. They had an economic interest in fostering international tension, as Marxists rightly emphasized — even if they may have overemphasized it in relation to ordinary public and political chauvinism and professional military ambition and insanity. That there is still an economic interest in arms and their development no minimally perceptive person will deny. But no multinational corporation can be suspected of promoting tension between the governments of countries in which it operates. IBM cannot be associated with any suspicion of stirring up trouble between France and Germany in order to sell computers, or ICI in order to sell chemicals. And it is persuasive that such suspicion in a suspicious world does not even arise. We should wish that there were more multinational operations between the Soviet Union and the West.

The multinational corporation also brings into existence the world's first truly effective international civil service — men and women who have a nominal loyalty to their country of origin, a rhetorical commitment to the country in which they serve and a primary loyalty to Shell, Du Pont, Philips, Nestlé or whatever company employs them. In a world that has suffered so much from national chauvinism, and especially in this century, this development is a small pacifying influence, a civilized step on from narrow, militant nationalism. But in the corporate rhetoric this dividend also goes completely unmentioned.

A wholly reliable tendency in the modern very large corporation, national or multinational, is for authority to pass from the stockholders or owners to the management. This means that it matters less and less, and eventually not at all, who ultimately owns the corporation or where it is owned, for the owners are without power. And as power passes to the management of the multinational company, it passes in part to the management of its various national entities. They have and, as a matter of course, must have a voice in the operations in which they are directly involved. While ownership of a multinational can be concentrated in the originating country, management, by its nature, must always be partly local.

This defense is also very rarely used. Failure of imagination may again be a factor. But many managements are unwilling to abandon the ridiculous fiction that their stockholders have power. All know that in the great multinational they are dispersed and passive, vote their proxies automatically for the management slate and are never otherwise heard from. But one must sustain the myth that the owners are still somehow important. The principal corporate folk rites, the stockholders' and directors' meetings, must be so scripted and staged as to give the impression that the owners are somehow a force in the affairs of the corporation.

In consequence, Canadians are not allowed to notice that the Canadian management of a U.S. company has, and must have, an extensive say in its Canadian operations. They are allowed to see only that the corporate shares are held mostly in the United States and to believe that this is the fact of importance. They are never persuaded to the view that, since those stockholders are totally without power, it makes little practical difference whether the shares are held by Canadians, Americans or Saudi Arabians.

Once again the approved mythology is damaging to multinational corporations; the truth is favorable. Nor would this truth have cost. Most stockholders know they are without power. To say and keep saying that the real power lies with management, that ownership is irrelevant and that extensive managerial autonomy must be granted to national entities as a matter of simple necessity would combine truth and plausibility with the calming of national fears.

Finally, critics allege that the multinational corporation exports jobs, capital and technology. This is one of the few matters on which the multinational enterprise has developed a partial defense. It holds that, in one way or another, it cares most about its home country and labor force — these are its primary interests. It should hold more often than it does that, as it goes abroad, others from abroad come in. The aggregate result is a more rapid spread of technology, a better international division of labor, greater productivity, greater aggregate employment. This is the old case for international trade.

It is an indication of the poverty of thought brought to these matters that even this traditional defense is so imperfectly made.

No one should assume from the foregoing that the multinational enterprise is without capacity for antisocial action. And its opportunities for such action are important, for they mark out the zones of danger that the leadership of any sensibly run enterprise would avoid.

The first of these zones of danger lies in the peculiar advantage held by the multinational enterprise in its relationship to the national state. The large corporation, national or multinational in its nature, has power in the state — a point I again emphasize. As a general rule, the foreign-based company brings its power to bear more tactfully than the domestically based multinational. Volkswagen could not risk the kind of lobbying in Washington on emission standards that is usual for General Motors. General Motors in Canada would not dream of instructing Ottawa as to company needs with the wonderful arrogance that Canadian Pacific once considered appropriate to its corporate dignity.

Though the multinational by its corporate nature must speak and work for its public needs, some things it must not do: in the face of unwanted regulation, environmental constraints or unpleasant labor relations, it must not threaten to move operations to another country. This threat exploits a particular advantage of multinationality; it will always, and reasonably, be considered an unfair exercise of such advantage.

The multinational must eschew bribery, even though this is an accepted source of political revenue or personal income in the country in question. There is, in all countries, something peculiarly pejorative about being in foreign pay. Bribery is especially unwise for an American multinational, for to be paid by Americans, who are thought to be rich and powerful and therefore ill-intentioned, is especially indiscreet. And only the hopelessly obtuse can now be unaware of the first truth concerning the United States, which is that there is no such thing as a secret in the American Republic, only varying lengths of time until it is revealed.

The multinational company must be extremely cautious about moving large funds in anticipation of the fall in a national currency or its appreciation. Such action also exploits an advantage peculiar to multinationality; and it helps precipitate the depreciation or appreciation against which the speculative action is taken. In an age of currency instability such action by large multinationals is certain to provoke attack.

Perhaps eventually, since injunctions to virtue are not often compelling in such matters, large international movements of funds by the multinationals will bring some kind of international registration and regulation. This I would welcome, as should all multinational executives whose primary concern is with less damaging ways of making money.

The multinational must also be cautious in matters concerning the environment. There is no reason to suppose that the foreign intruder is likely to be more damaging to ambient air, water, landscape or the tranquillity of life than the native enterprises. Industrial progress has, with rare exceptions, involved a movement from dirty processes to cleaner ones — from coal to oil and gas and from the filthy steam engine to the internal combustion engine and the electric motor. The modern problem of pollution arises not from the fact that processes have become dirtier but from the fact that, though cleaner, so much more is produced and consumed. Local industries, being generally older, will often be dirtier than the newer arrival from abroad with the newer process or product.

But no one should doubt that foreign dirt is worse than domestic dirt and also that a little new dirt (or noise or damage to the landscape) is worse than old dirt (or noise or damage) to which the senses have become adjusted. The multinational corporation must be acutely sensitive to community feelings on environmental issues. It should rarely if ever seek to override them. It should be scrupulous in conforming to existing laws and regulations and be the last to object to valid or seemingly valid new constraints. It should lose no opportunity to improve voluntarily on present practice. None of this is to get its executives a reward in heaven. It is to maximize their hope for peace on earth.

The most important precaution has to do with weapons and the arms trade. It cannot have escaped anyone's notice that, over the last several years, weapons producers, along with the oil multinationals, have attracted an overwhelming share of the criticism that has been directed at the international operations of corporations. All multinational firms have suffered for the errors and absurdities of Lockheed and Northrop.

The weapons business will always be regarded with unease, as well it should. No sensitive person can look with equanimity on investment in the instruments of mass death; in recent times much of this business has been with countries that have other and more pressing needs for their resources. Arms have absorbed money that could and should have gone to other and better purposes. And the fact that this business (like much oil business) is with governments and politicians adds appreciably to the likelihood of scandal. Payments to politicians are bribes; the same payments, when made to businessmen, are a finder's fee or commission.

So if a multinational enterprise wishes to steer clear of attack, it should steer clear of the arms trade (though one cannot have much hope of this advice being taken if the company is already in it). But most multinationals are not in this trade. And it might be stressed that Lockheed and Northrop and the weapons firms generally are not, in the strict sense, multinational corporations. Exceptionally for organizations of their size, they do not have a substantial overseas presence. (Much of their trouble came from the operations of footloose and irresponsible agents and operators.) In the recent scandals civilian manufacturing multinationals have not been much involved. An habitual but misguided clubbiness among businessmen has kept the civilian companies from emphasizing this fact. One way to avoid guilt by association would be to make clear that there is no association.

There are other matters on which the multinational enterprise, because it is in conflict with never completely latent national suspicion, must be cautious. But let me conclude with a word of summary. The literature of multinational enterprise is vast, and volume has been notably a substitute for good sense. Quite a bit has been written with a view to telling the multinationals what they would like to hear — or what they might like to pay to hear. Regularly it denies or elides the power that is evident to all seeing eyes. The conclusion from such a defense is that the multinational enterprise should not exist. And a large further part of the literature, the critical as distinct from the sycophantic half, accepts that it should not. This literature then omits to say what should take its place. The attack from the left on the multinationals is a curiously dead-end exercise.

My case is that the multinational should affirm the possession and exercise of power, accept that it must be responsibly employed, and then urge, as it is entitled to do, that there have been substantial advantages from such employment of power in the past. But it must also concede that there is opportunity for the exercise of power that is socially damaging. So, subject to negotiation, prohibition of abuses must be accepted if not always welcomed.

All this will require thought — a scarce and ill-regarded commodity in this area. I urge it nonetheless, not out of compassion, not out of fear for the future of the multinational enterprise, not even from any especial warmth of friendship but rather out of a sense of offended art. Great organizations which pride themselves on their performance in so many other matters should not be so outrageously obtuse in the case they make for themselves.

6

What Comes After General Motors

FOR MOST THINGS that are used, enjoyed or are otherwise in the service of man, the possibility of improvement is assumed. In technology and science a quest for something better is required. So, certainly, in the production of goods. So also, in principle at least, in government. To be sure, we are less confident than we once were that what is new is better. It may only be different and a handle for the salesman. And, as with DDT or the SST, there is a heightened suspicion that long-run damage may outweigh short-run miracles. But on one thing agreement remains. Man must keep trying. To say that anything artificed by humans — machine, consumer product, service, organization — is perfect and thus beyond further change is intolerable, a declaration of intellectual and possibly even moral bankruptcy.

There is, however, one exception. That is the modern large corporation. It is the ultimate work of God and man. In the United States belief in its final excellence is nearly absolute. There is a continuing discussion in economics departments and law schools as to whether it is too big and should be broken up. But all but the most pathologically optimistic or intransigently imperceptive recognize that this discussion is wholly liturgical. Nothing, it is known, will happen in practice. And there is, further, no discussion of how the corporation might be altered as to ownership or restructured as to management and public control.

In Europe matters are only marginally different. In Britain there are recurrent proposals from the left wing of the Labour Party for the nationalization of selected large corporations. These suggestions are accompanied by energetic, on occasion rather desperate, assurances from avowedly more responsible leaders that nothing in the way of practical action is really meant. In recent elections in France the left coalition offered a platform of nationalization for a considerable list of large corporations. This was widely regarded as reflecting an effort by the Communists to embarrass their socialist colleagues by insisting on impractical as well as unpopular action. There is active discussion in Europe and some affirmative steps to place representatives of workers on boards of directors. This, for reasons that will be noted presently, does not portend great change.

So General Motors is, indeed, thought to be the final work of God and man. That this is so is more than slightly strange, for GM and its corporate counterparts are known to displease a great many people. Their products invite less than universal acclaim. They are thought unresponsive to public need and convenience. Legally through their executives or illegally by various laundries, numerous corporations have been shown to have bought heavily into the recent Republican administration in the United States and to have invested, if less heavily, in some Democrats as well. Politicians and parties in foreign countries have been even more generously subsidized. This generosity has been much criticized, even though it is thought more wicked to receive than to give. In its pricing, procurement of materials, persuasion of consumers and distribution of the resulting income, the modern large corporation, numerous scholars now concede, functions increasingly as an independent force — as an instrument for the exercise of power in which there is responsibility primarily to itself. And this power is independent of, perhaps even above, the modern state.

Thus that the corporation should be exempt from the general and required instinct for change and improvement is surprising. One is led to wonder if one of the more subtle, but also one of the more vital, instruments of its power is the way in which it makes unfashionable, even vaguely eccentric or irresponsible, all discussion of how its basic constitution might, over time, be changed or reformed. I've often urged that this is one of the prime services of contemporary economic instruction. It presents the modern business firm as the powerless puppet of the market. So emasculated, why should it be the cause of any worry?

This, however, is an argument I need not again pursue. My purpose here is rather to raise the forbidden question: Were the modern large corporation not exempt from the general quest for improvement, along what lines would this proceed? I have, I must confess, little hope of progress. But, just possibly, we might get the discussion under way.

The central tendency of the modern large corporation — and the source of the problems which increasingly provoke discussion — can be quickly summarized: with time, increasing size and the increasing technical and social complexity of its task, it loses its legitimacy as an entrepreneurial and capitalist institution. It becomes instead an instrument of its own organization. There is even considerable technical agreement among students of the corporation as to what occurs. In the large, fully developed modern corporation the stockholder or owner is the functionless and powerless recipient of income and capital gains or, on occasion, of capital losses. He is represented nominally by a board of directors which is selected by the management and which, in one of the most predictable of political rites, then appoints the managers that appointed it. Token ownership of stock by the directors is required. But no member of the board of General Motors, the world's largest industrial firm, owes his position to the votes he commands. The last man of whom this could be said was Mr. Charles Stewart Mott, who died in February 1973. His entrepreneurial drive in pursuit of his property interest had by then receded a bit. He was ninety-seven.

Until a few years ago all directors of the Standard Oil Company of New Jersey, now Exxon — in 1977 the second largest in earnings and assets among industrial corporations in the world — were members of the top management hierarchy itself. They elected themselves to the board, which appointed them to their management posts. Such elections are hard to lose and such appointments hard to miss. More recently, this arrangement being so candid as to seem a trifle obscene, some outside directors have been selected. The selection, of course, was by the management.

Although the euthanasia of stockholder power has long been recognized as a basic tendency in the development of the great corporation — in the United States ever since the path-breaking work of Adolf A. Berle and Gardiner C. Means in the nineteen-thirties[1] — the reality is still thought embarrassing. So the board of directors is accorded ratification rights where expenditure of money is concerned; this is because action involving money, however symbolic, always conveys the impression of power. Members of the board are usually of mature years. This ensures that they will be given the respect that civilized men rightly accord to age or incipient senility. These gestures, together with the impressive solemnity that surrounds directorial proceedings, help suggest stockholder power. And important interests are served by this charade. The corporate management achieves an aspect of capitalist legitimacy. The comfort-seeking economist can continue to speak of "the owning and profit-maximizing entrepreneur" who is the pivot of his system and, in a more practical way, of the textbook market. The radical can keep the capitalist as an enemy, for the latter far more readily inspires antipathy than the faceless industrial

1 Adolf A. Berle and Gardiner C. Means, *The Modern Corporation and Private Property* (New York: Macmillan, 1933).

bureaucrat. But it is a charade that should mislead no one —
and, in fact, it misleads no one save the innocent and those with
a vested interest in the myth. The modern large corporation
is a powerful thing. And it is a creature in the service of its
own bureaucracy — of the massive organization that I have
elsewhere called its technostructure.

Having independent power and being the creature of its own
organization, the modern corporation, not surprisingly, serves
the purposes of its own management. These purposes are frequently different from those of the public or substantial parts
of it, and the latter are less than pleased. Specifically, the corporate bureaucracy, like all organization, seeks its own expansion. On such growth and aggrandizement, promotion and the
resulting salary, prestige and perquisites of the management
depend. Increasingly the persuasion of the customer and the
power of the state are brought to the support of this expansion.
Those who are not persuaded or who do not concur in this use
of the power of the state — who do not respond to the need
for more weapons, space exploration or highways and the supporting outlays that these require — deplore the result.

Also the expansion that rewards the industrial bureaucracy
proceeds at different rates in different industries — in automobiles more rapidly on occasion than in refining capacity, in
air conditioning more rapidly than in power supply, in advertising more rapidly than in newsprint. Nothing reliably relates
growth in one industry to growth in related and dependent
industries; thus we have fuel crises, power crises, newsprint
crises. And we find governments creating planning authorities
to do what the market no longer accomplishes. This has been
true even of Republican administrations in the United States,
although Republicans prefer to speak of czars rather than
planners. The word *czar* has an indubitably conservative ring,
and no one has told these worthy gentlemen what comes after
a czar.

Economic expansion has also come to be thought indifferent to environmental considerations; either industries or their products are thought negligent in their treatment of air, water, the countryside or our ears. That a bureaucracy empowered to pursue its own interest in growth should be hostile to such environmental effects accords also with expectations. So does the frequent claim of the auto industry that inflation could easily be cured by allowing a little more air pollution. In the modern large corporation truth tends to be what best serves economic interest.

We have also, in highly visible form, the effect on the distribution of income. There is first the distribution within the corporate hierarchy itself. As an individual rises in this hierarchy, his bureaucratic power increases — and among the things so increased is the power to set his own compensation. Thus compensation in the large corporation has become very generous. No one can seriously pretend (although some do) that it depends on the scarcity, and thus the market price, of the talent involved. It is the nature of organization that it takes men of average ability but diverse knowledge and combines their efforts for a result that is far better than any individual could achieve. To a marked extent organization provides a substitute for exceptional talent — something that is wholly recognized by the total indifference of the stock market to the changes in command in the great corporation and by the Stygian obscurity into which the greatest executive disappears on the day on which he retires. The name of no past president of General Motors, General Electric, General Dynamics or General Mills can be recalled by anyone except near neighbors, relatives and his wife.

Not only is compensation in the modern corporation unrelated to function, it can, on occasion, be inversely related to performance. The letters ITT are now a code reference to ill-considered and improper exercise of political influence. Political matters are peculiarly the responsibility of top management; on nothing, one would expect, could the man in charge be held more sternly to account. And this is especially so if the effect of such administration on earnings and securities values is adverse. For many recent years the manifold securities of ITT were among the least well regarded of portfolio assets. Mr. Harold Geneen, the head of the enterprise, was, all during this time, one of the three or four most highly paid executives in the United States. In 1973, the Chairman and chief executive of General Motors, Mr. R. C. Gerstenberg, received $923,000, and President Edward N. Cole, $833,000. This for both was an increase. The year is important, for it was the one in which these men failed to foresee the first fuel shortage and a shift to smaller cars. For the year ending January 1976, one of the highest paid executives in the United States was Meshulam Riklis. His conglomerate, Rapid-American, was one of the less profitable of its kind.

The salary of the chief executive of the large corporation is not a market reward for achievement. It is frequently in the nature of a warm personal gesture by the individual to himself. This no one likes to say. In all economic discussion we praise pecuniary motivation and consider it a legitimate as well as socially serviceable thing. But we are reluctant to suggest that the chief executives of the modern large corporation, when setting their own pay, are motivated to go after all the traffic will bear.

The compensation of the stockholder in the modern firm is no less anomalous. In the large corporation he is only rarely called upon for equity capital. Capital is all but exclusively provided out of earnings or by borrowing. The stockholder, we've seen, has no power and hence no role in the running of the firm. What then is the case for the private participating interest? None that wouldn't be as well served by converting his property right to a fixed return.

Associated with the position of the stockholder is the increasingly asymmetrical role of profits and losses in the modern large corporation. Profits may still be made. Their function is still much celebrated. So are the virtues of private enterprise. And so are its immunities — its right to be free from government interference. All this changes dramatically when the great corporation sustains a loss. Then it becomes too large and too important to be allowed to fail and go out of business. So long as they were making profits, the Penn Central, Lockheed and the Franklin National Bank were flagships of the private enterprise system. Once their profits turned into losses, they became the highly deserving wards of the state. A few years back Consolidated Edison, which prudently calls itself an "investor-owned utility," entered upon a particularly dark period in a generally dark life. The *New York Times* ran a question-and-answer series on its probable fate. Would it go bankrupt? "No, the ramifications of such action are too severe." Would it be taken over by the public? "Yes, for investors this might be the better way." This, it was agreed, would invite "a major ideological change." It was implied that, as an alternative to bankruptcy, this ideological change could be accomplished with relative ease.

As in the United States, so in other countries. In Britain, there is strong objection to public ownership except as in the case of Rolls-Royce, British Leyland, the shipyards or the earlier instance of the coal pits, where private mismanagement or miscalculation brought disaster. Then it is favored. The Italians have built a large public sector of the economy from the failures of private enterprise and the French a slightly smaller one. The tendency is the same elsewhere. When considering what comes after General Motors, we cannot exclude some considerable role for the government. The public and private embrace is already great.

The first change in the modern large corporation should involve, plausibly, the two bodies that have lost function — the stockholders and the board of directors.

The case for private ownership diminishes drastically when the stockholder ceases to have power — when he or she or it becomes a purely passive recipient of income. Then, since the management is a self-governing, self-perpetuating bureaucracy, no claim can be made to the traditional immunity associated with property ownership. A logical course will be for the state to replace the helpless stockholder with an effective supervisory and policy-setting body. One forthright way to accomplish this would be to have a public holding company take over the common stock. There is, of course, no case for singling out one class of property for sequestration. The common stock would be valued, perhaps by reference to past stock market prices or perhaps by more specific appraisal of assets, and the stockholder paid off with fixed, interest-bearing securities. Whether these should be guaranteed by the state is an interesting question. It is possible that they should be; this would accept as fact the public rescue of creditors that is now normal. It should be noted that no public expenditure would be involved in this transaction, only an exchange of assets.

Were the criterion for this action that the firm be a full, self-governing bureaucracy — that the power of the management be plenary and that of the stockholder nil — some hundred or more of the largest industrial firms in the United States would be eligible, together with a fair number of the larger utility, transportation, financial and merchandising corporations. In practice, as a man of recognized caution, I would be content to start with three obvious categories. The first would be the big weapons firms — Lockheed, General Dynamics, Thiokol — which are now, except in name, public enterprises. The government is overwhelmingly their most important customer; it supplies them with most of their working and much of their fixed capital, covers their cost overruns and otherwise stands by to rescue them when they are in trouble. What is called private enterprise is here a disguise for costly and unexamined access to the public trough. Then there are the oil majors. These have a strong environmental impact, a large influence on foreign policy and a stranglehold on the public pocketbook. Public ownership in this field has worked well elsewhere, notably in Britain, Austria and numerous other countries. It would, no one should doubt and least of all the oil men, be widely popular. And the earnings would offset the losses from the third and inescapable category, which, of course, is the losers — the railroads and the other yet undisclosed candidates for public assistance as they come on hard times. It is now taken for granted that the state will inherit the turkeys. And this is necessary; we can't get along without railroads, banks or perhaps even the Chrysler Corporation, which has already required rescue in its British manifestation. But, this being so, it is surely proper that the public should also inherit some profit-makers. There is a case for symmetry of a sort here.

As equity passes to the public, so would capital gains. Thus would be eliminated a wholly adventitious source of public enrichment derived extensively from the accident of parentage or grandparentage. The blight that was always over the careers of men like Nelson Rockefeller would thus, in time, be removed.

However, it is important that liberals not have false hopes of the effect of the change on operational detail. One reason the private stockholder was excluded from management was that his intrusion, since it was uninformed, was damaging. So it is with any outsider; a public owner cannot and should not participate in the routine of management. This has come to be well understood in many countries, with the consequence that public ownership has been reconciled with extensive managerial autonomy and marked commercial success. The example of automobile manufacturers — Renault, Alfa-Romeo, Volkswagen — is especially impressive.

Where the private stockholder disappears, the board of directors should be replaced with what might variously be called a board of public auditors or a board of public inspectors. I propose to use the first name; it is important only that the designation carry no connotation of operational authority. This board should be of moderate size — perhaps eight members. A minority of the members, perhaps three, should be selected, as now, by the management. This minority would include the senior executives of the firm. The remaining five would be designated as public members by the state. Like judges, they should be men and women of quality and strong public instinct. They should not be of excessive age; they should be of persistent, informed and disciplined curiosity. They would be expected to take their jobs seriously, meet at least monthly, and they would be paid. Since the board of public auditors would have no operational duties, I do not see it as a useful place for the exercise of trade union or other specific participant interest. Nor do I think that participation on present boards of directors, which are operationally powerless bodies, much serves trade union ends. Here also the defense of the worker interest is more effectively served by traditional union means. That the board of public auditors would reflect the consumer interest is inherent in its public character. The consumer *is* the public and vice versa.

The board would ratify the selection and promotion of top management and, in the event of self-perpetuating mediocrity, replace it. And, in addition to the conventional financial audit, it would maintain continuing surveillance to ensure honest conformity with legally established public goals on bribery, safety of products, pollution, political neutrality and the rest. To this end, it should, of course, have full access to the information available in the firm — to prices, costs, investment planning, product design, advertising and merchandising methods and plans, and much more. On important or continuing matters it would function by committee and could request staff assistance.

The board of auditors would also generally be the place for informed discussion of the public impact of the major policies of the firm. Some public benefit would come from this discussion and the resulting response. Policy directives on public matters would, of course, be binding on management. It would be important, however, that this power be used sparingly. Management must retain the major powers of decision and therewith the capacity to perform. And it must be held responsible for effective performance. It is a sound rule that matters of public urgency — those pertaining to the environment, product design, advertising — should be controlled by general law, not by individual company decision.

Both the findings and deliberations of the board of public auditors should be known. Any notion that the great corporation is a private entity is obsolete. From the reaction to what is made available by the board of public auditors would come some of the knowledge that would guide its discussions and findings. From this information and discussion would also come the raw material for legislation setting and refining the rules on what can be done to air, water, the tranquillity of life, the safety of consumers and by way of public bamboozlement — in short, for establishing the parameters within which corporate growth and profit-making should proceed. These parameters, all conservatives should recognize, are increasingly their protection against assault on the idea of growth itself.

Plenary power should, obviously, be accorded to the board of public auditors for setting the range of top executive compensation. That power is now exercised by the people who receive the compensation themselves or by their appointees. This is far too convenient. While to reduce executive compensation in the new public corporations to civil service levels might be too radical a step, there is no excuse for the present munificence, and there is no relation to incentives. That is because no executive can possibly have it suggested that he adjusts his effort or the exercise of his intelligence to his compensation.

If large tasks are to be performed, large corporations are necessary. In considering what comes after General Motors the task is to bring the large organization into line with the modern reality. As it develops, stockholder power becomes irrelevant, the stockholder representation by the board of directors a farce. The corporation becomes a self-governing and self-perpetuating instrument of its management, and this the myth no longer conceals. By ridding it of the obsolete features, substituting a public surveillance and at the same time safeguarding its operational autonomy, the chances for the orderly survival of the great corporation could well be strengthened. What I here suggest, modified no doubt by further discussion, could well be a goal of all enlightened executives. Of this, however, I have no great hope. The most one can ask is that a few rational professionals here and there will be repelled by the myth of stockholder power and directorial omniscience that they are now required to perpetrate and perpetuate and will accept that the corporation is unfinished business, that there might be room for further change.

7

The Founding Faith:
Adam Smith's *Wealth of Nations*[1]

> It is not from the benevolence of the butcher, the brewer, or
> the baker, that we expect our dinner but from their regard to
> their own interest. We address ourselves, not to their humanity
> but to their self-love, and never talk to them of our own neces-
> sities but of their advantages.
>
> *Wealth of Nations*

ADAM SMITH, not to put too fine an edge on matters, was Scotland's greatest son. *Wealth of Nations* is his greatest and almost
his only book. As Karl Marx is much too valuable a source of
social insight to be left as the exclusive property of the Communists, so Adam Smith is far too wise and amusing to be
relegated to conservatives, few of whom have ever read him.

Smith was born in 1723 in what was then the small port town
of Kirkcaldy on the Firth of Forth across from Edinburgh. The
enduring exponent of the freedom of trade was the son of the
local collector of customs. After study at the evidently excellent
local school, he went on to the University of Glasgow and then
to Oxford (Balliol College) for six years. Returning to Scotland, he became, first, professor of logic and then, in 1752, pro-

[1] This is a revision of a paper given in Kirkcaldy, Scotland, Adam Smith's birthplace, in June of 1973, at a gathering to celebrate the 250th anniversary of his
birth. In *The Age of Uncertainty* (Boston: Houghton Mifflin, 1977) I have dealt
at more length (and with some overlapping) with the events of Smith's life.

fessor of moral philosophy at the University of Glasgow. This chair he resigned in 1763 to travel on the Continent as the well-paid tutor of the young Duke of Buccleuch, a family possessed to this day of a vast acreage of dubious land on the Border. In Europe Smith made the acquaintance of the Physiocratic philosophers and economists Quesnay and Turgot, as well as Voltaire and other notable contemporaries, and used his time and mind well. He then returned to Kirkcaldy where, for the next ten years, subject to lengthy sojourns in London and to the despair of some of his friends who feared he would never finish, he engaged himself in the writing of *Wealth of Nations*.

This book was published in 1776, a few weeks before the Declaration of Independence, and there was some relationship between the two events. Unlike his friend David Hume (who died that August) and consistently with his economic views, Smith deplored the separation. He had wanted instead full union, full and equal representation of the erstwhile colonies in Parliament, free trade within the union, equal taxation along with equal representation and the prospect that, as the American part developed in wealth and population, the capital would be removed from London to some new Constantinople in the West.

Wealth of Nations, at least among the knowledgeable, was an immediate success. Gibbon wrote: "What an excellent work is that with which our common friend Mr. Adam Smith has enriched the public . . . most profound ideas expressed in the most perspicuous language."[2] Hume, in a much quoted letter, was exuberant:

Euge! Belle! Dear Mr. Smith. — I am much pleased with your performance, and the perusal of it has taken me from a state of great anxiety. It was a work of so much expectation, by yourself, by

2 Edward Gibbon, quoted in John Rae, *Life of Adam Smith* (New York: Augustus M. Kelley, 1965), p. 287.

your friends, and by the public, that I trembled for its appearance, but am now much relieved . . . it has depth and solidity and acuteness, and is so much illustrated by curious facts that it must at last attract the public attention.[3]

The public response — to two volumes costing £1 16s., the equivalent of perhaps forty dollars today — was also good. The first edition was soon sold out, although this intelligence would be more valuable were the size of the edition known. Smith spent the next couple of years in London, being, one gathers, much feted by his contemporaries for his accomplishment, and then, having been appointed Commissioner of Customs in Edinburgh, an admirable sinecure, he returned to Scotland. He died there in 1790.

By this time, *Wealth of Nations,* though at first ignored by politicians, was having an influence on men of affairs. A year and a half after Smith's death, Pitt, introducing his budget, said of Smith that his "extensive knowledge of detail and depth of philosophical research will, I believe, furnish the best solution of every question connected with the history of commerce and with the system of political economy."[4] Not since, in the non-socialist world at least, has a politician committed himself so courageously to an economist.

Smith has not been a popular subject for biographers. He was a bachelor. His best-remembered personal trait was his absent-mindedness. Once, according to legend, he fell into deep thought and walked fifteen miles in his dressing gown before regaining consciousness. His manuscripts, by his instruction, were destroyed at his death. He disliked writing letters, and few of these have survived. The papers of those with whom he did correspond or which reflected his influence were destroyed, mostly because of lack of interest, and some, it appears, as late as 1941 or 1942. Adam Smith's only other major published

---

[3] David Hume, quoted in Rae, p. 286.
[4] William Pitt, before the House of Commons on February 17, 1792, quoted in Rae, pp. 290–291.

work, *The Theory of Moral Sentiments,* reflects interests that were antecedent to his concern with political economy. No biography of Adam Smith has superseded that by John Rae, first published nearly eighty years ago.

Though Smith's life has attracted little attention, much has centered on *Inquiry into the Nature and Causes of the Wealth of Nations,* to give the title of his masterpiece its full resonance. With *Das Kapital* and the Bible, *Wealth of Nations* enjoys the distinction of being one of the three books that people may refer to at will without feeling they should have read them. Scholarly dispute over what is Smith's principal contribution has continued ever since publication. This is partly because there is so much in the book that every reader has full opportunity to exercise his own preference.

Exercising that preference, I have always thought that two of Smith's achievements have been neglected. One, mentioned by Gibbon, is his gift for language. Few writers ever, and certainly no economist since, have been as amusing, lucid or resourceful — or on occasion as devastating. Many people rightly remember his conclusion that "People of the same trade seldom meet together, even for merriment and diversion, but the conversation ends in a conspiracy against the public, or in some contrivance to raise prices."[5] There are many more such gems. He noted that "The late resolution of the Quakers in Pennsylvania, to set at liberty all their negro slaves, may satisfy us that their number cannot be very great."[6] And, anticipating Thorstein Veblen, he observed that "With the greater part of rich people, the chief enjoyment of riches consists in the parade of riches . . ."[7] On the function or nonfunction of stockholders, no one in the next two centuries was more penetrating in however many words: "[Stockholders] seldom pretend to understand

5 Adam Smith, *Wealth of Nations* (Edinburgh: Adam and Charles Black, 1863), p. 59.
6 Smith, p. 172.
7 Smith, p. 79.

any thing of the business of the company; and when the spirit
of faction happens not to prevail among them, give themselves
no trouble about it, but receive contentedly such half-yearly or
yearly dividend as the directors think proper to make to
them."[8] One of Smith's most useful observations, which should
always be kept in mind when alarm substitutes for thought, is
not in *Wealth of Nations*. On hearing from Sir John Sinclair
in October 1777 that Burgoyne had surrendered at Saratoga
and of his friend's fear that the nation was ruined, Smith said,
"There is a great deal of ruin in a nation."[9]

Also neglected now are the "curious facts" that enchanted
Hume and of which *Wealth of Nations* is a treasure house.
Their intrusion has, in fact, been deplored. As a writer Smith
was a superb carpenter but a poor architect. The facts appear
in lengthy digressions which have been criticized as such. But
for any discriminating reader it is worth the interruption to
learn that the expenses of the civil government of the Massachusetts Bay Colony "before the commencement of the present
disturbances," meaning the Revolution, were only £18,000 a
year and that this was a rather sizeable sum compared with the
expenses of New York and Pennsylvania at £4500 each and of
New Jersey at £1200. (These and numerous other details on
the colonies reflect an interest which John Rae believes was
stimulated by Benjamin Franklin.)

Also, were it not for Smith, we might not know that after
a bad storm, or "inundation," the citizens of the Swiss canton
of Underwald (Unterwalden) came together in an assembly
where each publicly confessed his wealth to the multitude and
was then assessed, *pro rata,* for the repair of the damage. Or
that, at least by Smith's exceptionally precise calculation,
Isocrates earned £3333 6s, 8d. (upward of 60,000 of today's dollars) for "what we would call one course of lectures, a number
which will not appear extraordinary, from so great a city to

8 Smith, p. 333.
9 Adam Smith, quoted in Rae, p. 343.

so famous a teacher, who taught too what was at that time the most fashionable of all sciences, rhetoric."[10] Or that Plutarch was paid the same. Or, continuing with professors, that those who are subject to reward unrelated to their capacity to attract students will perform their duty in "as careless and slovenly a manner as that authority will permit" and that in "the university of Oxford, the greater part of the public professors [those with endowed or salaried chairs] have, for these many years, given up altogether even the pretence of teaching."[11]

So no one should neglect Smith's contribution to expository prose and "curious facts." Now as to economic thought and policy. Here a sharp and obvious distinction must be made between what was important in 1776 and what is important now. The first is very great; the second, save in the imagination of those who misuse Smith as a prophet of reaction, is much less so. The business corporation which Smith deplored and the wealth that accumulated in consequence of his advice combined to reduce the later relevance of that advice. But first we must consider his meaning in 1776.

Smith's economic contribution to his own time can be thought of as falling into three categories — method, system and advice. The second, overflowing onto the third, is by far the most important.

As to method, Smith gave to political economy, later to become economics, the basic structure that was to survive almost intact at least for the next hundred and fifty years. This structure begins with the problem of value — how prices are set. Then comes the question of how the proceeds are shared — how the participants in production are rewarded. These latter are the great trinity of labor, capital and land. Along the way is the role of money. Thereafter come banking, international trade, taxation, public works, defense and the other functions of the

10 Smith, p. 61.
11 Smith, p. 342.

state. Other writers, notably the Physiocrats, had previously given political economy a fairly systematic frame, although, as Alexander Gray, a later Glasgow professor, observed, they had "embellished it with strange frills." But it was Smith who, for the English-speaking world, provided the enduring structure.

The framework, in turn, was more important than what it enclosed. Although Smith's treatment of value, wages, profits and rents suggested what was to follow, it was, in all respects, a beginning and not an end. Thus, as one example, Smith held that the supply of workers would increase, *pari passu*, with an increase in the sustenance available for their support. David Ricardo translated this thought into the iron law of wages — the rule that wages would tend always to fall to the bare minimum necessary to sustain life. And Thomas Robert Malthus, going a step further, adduced his immortal conclusion that people everywhere would proliferate to the point of starvation. Subsequent scholars — the marginal utility theorists, Alfred Marshall, others — added further modifications to the theory of prices, wages, interest, profits and rent, and yet further transmutations were, of course, to follow. Smith was left far behind.

But the structure he gave to economics and the explanation of economic behavior that it contained were, for Smith, only steps in the creation of his larger system — his complete view of how economic life should be arranged and governed. This was his central achievement. It provided a set of guiding rules for economic policy that were comprehensive and consistent without being arbitrary or dogmatic.

In the Smithian system the individual, suitably educated, is left free to pursue his own interest. In doing so, he serves not perfectly but better than by any alternative arrangement the common public purpose. Self-interest or selfishness guides men, as though by the influence of "an invisible hand," to the exercise of the diligence and intelligence that maximize productive effort and thus the public good. Private vice becomes a public virtue, which has been considered ever since a most convenient thing.

In pursuit of private interest, producers exploit the opportunities inherent in the division of labor — in, broadly speaking, the specialized development of skill for the performance of each small part of a total task of production. Combined with the division of labor is the natural propensity of man "to truck, barter or exchange." The freedom of the individual to do his best both in production and in exchange is inhibited by regulation and taxation. Thus the hand of the state should weigh on him as lightly as possible. The limiting factor on the division of labor — roughly, the scale of specialized productive activity — is the size of the market. Obviously this should be as wide as possible. Thus Smith's case against internal monopolistic or international restrictions on trade.

Smith's precursors, the mercantilists, held that national well-being and national strength derived from and required the accumulation by the country of precious metal. Smith held — as one would now say and as he in effect did say — that they derived instead from the productivity of the labor force. Given an industrious and productive labor force, in the most majestic of Smith's arguments, there would be no need to worry about the stock of gold. The gold would always come.

Such, in greatest compression, is the Smithian system — the one that Pitt proclaimed as "the best solution of every question connected . . . with the system of political economy."

Smith's third contribution was in the field of practical policy. His advice — on banking, education, colonies, support of the sovereign (including his famous canons of taxation and extending even to recommendations for the reform of taxation in France), public works, joint-stock companies and agriculture — was infinitely abundant. No economist since has offered more. With many exceptions and frequent modifications to fit the circumstances, this advice is in keeping with Smith's system. The bias in favor of freeing or unburdening the individual to pursue his own interest is omnipresent, and so is Smith's belief that men will toil effectively only in the pursuit of pecuniary self-interest. There will be occasion for a further word on this advice; now we must see what of Smith survives.

Needless to say, the mordant language and the curious facts remain; it is too bad they are not more read and enjoyed. Also Smith's concept of the economic problem and the division of the subject between value and distribution are still to be found in that part of the textbooks that economists call microeconomics. His particular conclusions as to how prices, wages, rents and return to capital are determined and his views on gold, paper currency, banks and the like are now mostly of antiquarian interest.

Nor does all of the abundant advice just mentioned have modern meaning. It better illuminates life in the eighteenth century than it does any current problems. Until recently the textbooks on taxation included reverent mention of Smith's four great canons. But no one now coming to them without knowledge of their author would think them very remarkable. That taxes should be certain or predictable and arbitrary in their bite; that they should be so levied and collected as to fit the reasonable convenience of the taxpayer; and that the cost of collection should be a modest part of the total take were all important in 1776. They still are, but these things are fairly well accepted now.

Smith's fourth canon, that the "subjects of every state ought to contribute towards the support of the government, as nearly as possible, in proportion to their respective abilities; that is, in proportion to the revenue which they respectively enjoy under the protection of the state,"[12] could be taken as a prescription for a proportional (i.e., fixed percentage) as distinct

12 Smith, p. 371.

from a progressive income tax. Some beleaguered rich have so argued. In fact, Smith was speaking only of what seemed possible and sensible in his own day. He would, almost certainly, have moved with the times. It might be added that his modest prescription gives no place to tax shelters, special treatment of state and municipal bonds or the oil depletion allowance and no comfort to those who otherwise believe that they were intended by nature to be untroubled by the IRS. Numerous of the great rich in the United States would find even Adam Smith's proportional prescription rather costly as compared with what they now pay.

The next and more interesting question concerns Smith's system — his rules for guiding economic life. What of that survives? Is economic life still directed in appreciable measure by the invisible hand — in modern language, by the market? What has happened to the notion of the minimal state, and is it forever dead? And what of Smith's plea for the widest possible market both within and between nations?

Nothing so rejoices the conservative soul as the thought that it all survives. It doesn't. Smith was the victim of one major miscalculation. And, as earlier noted, he was damaged by the institution that he deplored, the business corporation. His system was also gravely impaired by the very success of the prescription that he offered.

Smith's miscalculation was of man's capacity, perhaps with some social conditioning, for cooperation. He thought it negligible. Men would work assiduously for their own pecuniary advantage; on shared tasks, even for shared reward, they would continue to do as little as authority allowed. Only in defeating or circumventing that authority — in minimizing physical and intellectual toil and maximizing indolence and sloth — would they bring real effort and ingenuity to bear. But not otherwise. People work only if working for themselves; there is no more persistent theme in *Wealth of Nations.* It is why government

tasks are poorly performed. It is why civil servants are an un-
civil and feckless crew. It is his case against the British bureaucracy in India. It is why the Oxford professors on a secure salary
lapsed into idleness. And especially it is why, in Smith's view,
joint-stock companies, except for routine tasks, had little to
commend them. Their best chance for survival, one to which
the minds of the directors almost invariably turned, was to
obtain a monopoly of their industry or trade, a tendency to
which Smith devoted some of his finest scorn. Otherwise their
officers concerned themselves not with enriching the company
but with enriching themselves or not enriching anyone.

In fact, experience since Smith has shown that man's capacity
for cooperative effort is very great. Perhaps this was the product
of education and social conditioning, something that no one
writing in the eighteenth century could have foreseen. Conceivably Smith, handicapped by his environment, judged all
races by the Scotch (as we are correctly called) and their much
celebrated tendency to self-seeking recalcitrance. Most likely he
failed to see the pride people could have in their organization,
their desire for the good opinion or esteem of their co-workers
and their satisfaction in what Thorstein Veblen would call "the
instinct of workmanship."

In any case, governments in the performance of public tasks,
some of great technical and military complexity, corporations
in pursuit of growth, profit and power, and advanced socialist
states in pursuit of national development and power have been
able to enlist a great intensity of cooperative effort.

The most spectacular example of this cooperative effort —
or perhaps, to speak more precisely, of a successful marriage of
cooperative and self-serving endeavor — has been the corporation. The way in which the corporation has come to dominate
economic life since 1776 need hardly be emphasized. This development Smith did not foresee, an understandable flaw. But
he also thought both the form or structure and the cooperative

or organized effort inherent in corporate development flatly impossible.

The corporation that Smith did not think possible was then extensively destructive of the minimal state that he prescribed. This destruction it accomplished in several ways. The corporation had needs — franchises, rights-of-way, capital, qualified manpower, support for expensive technological development, highways for its motor cars, airways for its airplanes — which only the state could supply. A state that served its corporations as they required quickly ceased, except in the minds of more romantic conservatives, to be minimal.

Also, a less evident point, the economy of which the great corporation was so prominent a part along with the unions was no longer stable. The corporation retained earnings for investment, as did the individuals it enriched. There was no certainty that all of such savings would be invested. The resulting shortage of demand could be cumulative. And in the reverse circumstances wages and prices might force each other up to produce an enduring and cumulative inflation. The state would be called upon to offset the tendency to recession by offsetting excess savings and stabilizing the demand for goods. This was the message of Keynes. And the state would need to intervene to stabilize prices and wages if inflation were to be kept within tolerable limits. Both actions, traceable directly to corporate and counterpart union development, were blows at the Smithian state.

The corporation, as it became very large, also ceased to be subordinate to the market. It fixed prices, sought out supplies, influenced consumers and otherwise exercised power not different in kind from the power of the state itself. As Smith would have foreseen, this power was exercised in the interest of its possessors, and on numerous matters — the use of air, water and land — the corporate interest diverged from the public interest. It also diverged where, as in the case of the weapons firms, the corporation was able to persuade the state to be its customer. Corporate interest did not coincide with public interest as the Smithian system assumed. So there were appeals by the public to the government for redress and further enhancement of the state. All this was not as Smith would have thought.

Finally, Smith's system was destroyed by its own success. In the nineteenth century and with a rather deliberate recognition of their source, Britain was governed by Smith's ideas. So, though more by instinct than by deliberate philosophical commitment, was the United States. And directly or through such great disciples as the French economist J. B. Say, Smith's influence extended to Western Europe. In the context of time and place, the Smithian system worked; there was a vast release of productive energy, a great increase in wealth, a large though highly uneven improvement in living standards. Then came the corporation with its superior access to capital (including that reserved from its own earnings), its great ability to adapt science and technology to its purposes and its strong commitment to its own growth through expanding sales and output. This, and by a new order of magnitude, added to the increase in output, income and consumption.

This increase in well-being was also damaging to the Smithian system. It was not possible to combine a highly productive economy and the resulting affluence with a minimal state. Public regulation had to develop in step with private consumption; public services must bear some reasonable relationship to the supply of private services and goods. Both points are accepted in practice, if not in principle. A country cannot have a high consumption of automobiles, alcohol, transportation, communications or even cosmetics without public rules governing their use and public facilities to rescue people from accidents and exploitation. The greater the wealth, the more men need to protect it, and the more that is required to pick up the discarded containers in which so much of it comes. Also in rough accord with increased private consumption goes an increased demand for public services — for education, health care, parks and public recreation, postal services and the infinity of other things that must be provided or are best provided by the state.

Among numerous conservatives there is still a conviction that the minimal state was deliberately destroyed by socialists, planners, *étatists* and other wicked men who did not know what they were about, or knew all too well. Far more of the responsibility lies with Smith himself. Along with the corporation, his system created the wealth that made his state impossible.

In one last area, it will be insisted, Adam Smith does survive. Men still respect his inspired and inspiring call for the widest possible market, one that will facilitate in the greatest degree the division of labor. After two centuries the dominant body of opinion in industrial nations resists tariffs and quotas. And in Europe the nation-states have created the ultimate monument to Adam Smith, the European Economic Community. In even more specific tribute to Smith, it is usually called the Common Market.

Even here, however, there is less of Smith than meets the eye. Since the eighteenth century, and especially in the last fifty years, domestic markets have grown enormously. That of insular Britain today is far greater than that of Imperial Britain at the height of empire. The technical opportunities in large-scale production have developed significantly since 1776. But national markets have developed much, much more. Proof lies in the fact that General Motors, IBM, Shell and Nestlé do not produce in ever larger plants as would be the case if they needed to realize the full opportunities inherent in the division of labor. Rather, they regularly produce the same items in many relatively small plants all over the world. Except perhaps in the very small industrial countries — Holland, Belgium, Luxembourg — domestic markets have long been large enough so that even were producers confined to the home market, they would realize the full economies of scale and the full technical advantages of the division of labor.

The Common Market and the modern enlightenment on international trade owe much more to the nontechnical needs of the modern multinational corporation than they do to Adam Smith. The multinational corporation stands astride national boundaries.[13] Instead of seeking tariff support from the state against countries that have a comparative advantage, it can move into the advantaged countries to produce what it needs. At the same time modern marketing techniques require that it be able to follow its products into other countries to persuade consumers and governments and, in concert with other producers, to avoid the price competition that would be disastrous for all. So, for the multinational corporation, tariffs, to speak loosely and generally, are both unnecessary and a nuisance. It would not have escaped the attention of Adam Smith, although it has escaped the attention of many in these last few years, that where there are no corporations, as in agriculture, the Common Market is more contentious and less than popular. The tariff enlightenment following World War II has resulted not from a belated reading of *Wealth of Nations* but from the much more powerful tendency for what serves the needs of large enterprises to become sound public policy.

But if time and the revolution that he helped set in motion have overtaken Smith's system and Smith's advice, there is one further respect in which he remains wonderfully relevant. That is in the example he sets for professional economists — for what, at the moment, is a troubled, rather saddened discipline. Smith is not a prophet for our time, but, as we have seen, he was magnificently in touch with his own time. He broke with the mercantilist orthodoxy to bring economic ideas abreast of the industrial and agricultural changes that were only then

---

[13] See the earlier essay, "The Multinational Corporation: How to Put Your Worst Foot Forward or in Your Mouth."

just visible on the horizon. His writing in relation to the Industrial Revolution involved both prophecy and self-fulfilling
prophecy. He sensed, even if he did not fully see, what was
about to come, and he greatly helped to make it come.

The instinct of the economist, now as never before, is to
remain with the past. On that, there is a doctrine, a theory —
one that is now elaborately refined. And there are practical
advantages. An economist's capital, as I've elsewhere observed,
lies in what he knows. To stay with what is accepted is also
consistent with the good life — with the fur-lined comfort of
the daily routine between suburb, classroom and office.[14] To
such blandishments, economists are no more immune than other
people. The tragedy lies in their own resulting obsolescence.
As the economic world changes, that proceeds relentlessly, and
it is a painful thing.

Remarkably, the same institution, the corporation, which
helped to take the economic world away from Adam Smith, has
taken it away from the mature generation of present-day economists. As even economists in their nonprofessional life concede, the modern corporation controls prices and costs, organizes
suppliers, persuades consumers, guides the Pentagon, shapes
public opinion, buys politicians and is otherwise a dominant
influence in the state. In its contemporary and comprehensively
powerful form it also, alas, figures only marginally in the accepted economic theory. That theory still holds the business
firm to be solely subordinate to the market, solely subject to
the authority of the state and ultimately the passive servant of
the sovereign citizen. None of this being so, scholars have lost
touch with reality. Older economists and some younger ones
are left only with the hope that they can somehow consolidate
their forces and live out the threat. It is a fate that calls less for
criticism than for compassion.

It is not a fate that Adam Smith would have suffered. Given

---

[14] See the earlier essay, "Economists and the Economics of Professional Contentment."

his avid empiricism, his deep commitment to reality and his profound concern for practical reform, he would have made the modern corporation and its power and the related power of the unions and the state an integral part of his theoretical system. His problem would have been different. With his contempt for theoretical pretense and his intense interest in practical questions, he might have had trouble getting tenure in a first-rate modern university.

8

Defenders of the Faith, I:
William Simon

WHILE I AM NOT ACQUAINTED WITH Mr. William Simon, I don't believe that I would be his first choice to review his defense of the free enterprise system, *A Time for Truth*.[1] The former energy "czar" and Secretary of the Treasury here urges business-men to give their money to people and institutions of which he approves. He told a group of college administrators a while back that I was more or less precisely the kind of person he did not have in mind. It is obvious that I must lean over backwards to be judicious and fair. But happily that is, in any case, my tendency.

In this spirit I would like at the outset to say that Mr. Simon has a strong point; this, like all other occasions, *is* a good time for truth. In economics, unfortunately, one person's truth is another's fallacy (although with this Mr. Simon does not agree), so, to ensure fairness on the subject, I propose to leave eco-nomic questions aside, at least for the time being.

I must also leave aside the question of whether Mr. Simon is as secure in his views as he asserts. A truly secure author doesn't usually get someone else to write a preface as a crutch. I find it disturbing that Mr. Simon uses two such prostheses,

[1] Preface by Milton Friedman. Foreword by F. A. Hayek. (New York: Reader's Digest Press, McGraw-Hill, 1978.)

one from Milton Friedman, another from F. A. Hayek. One thinks of a man negotiating a personal loan who gets an endorsement from both Walter Wriston and David Rockefeller. I do note that Professor Friedman calls this case for capitalism "brilliant and passionate . . . by a brilliant and passionate man . . . a profound analysis of the suicidal course on which our beloved country is proceeding . . ." Perhaps Mr. Simon found that this sort of praise is hard to reject, although it could also jeopardize Professor Friedman's reputation for careful, restrained, scientific statement. However, he may have thought that overstatement would not be much noticed in this volume.

I must leave aside as well a yet more objective conflict with truth, which is whether Mr. Simon is truthful when he says he wrote this document. Others have held that it was written by Ms. Edith Efron, who here gets credit only for assistance from "conception to execution [sic]." But this will be a matter for later attention. With some others I am planning a new organization to require all statesmen, orators and Washington malefactors and their publishers to reveal the true source of their prose. This force for truth will be called SAD — Society for Authorship Disclosure.

Finally, all devotees of the truth will wish Mr. Simon had been less discriminating in its use. He is commendably frank in his political likes and dislikes. He despises all Democrats, most Republicans, all congressmen, almost all newspapermen, many businessmen and bankers, and especially he despises bureaucrats. (Apart from Professors Friedman and Hayek and Mr. William Buckley, there aren't too many people he does not despise.) But he tells here of going to Moscow in April 1975 for a worthy but modest public relations exercise on behalf of Soviet-American trade. What he calls "a staff group from the departments of State and Commerce; and others" went along. It was a relief, he tells, to be on the way home, and as they soared out of the Soviet capital on Air Force Two, "seventy-eight dignified representatives of the United States of America shouted and applauded like youngsters in sheer relief..." A more candid man would condemn taking seventy-eight people all that distance for such a job. That, in all truth, Mr. Simon, was bureaucracy run wild.

There has also been complaint that the author is a trifle disingenuous (truth again) in the treatment of his relationship to the great New York City financial disasters of the mid-seventies. His indignation over the hocus-pocus and concealment in New York financial management is unlimited. But he is himself a certified and experienced municipal bond expert, and during the worst of the thimblerigging and prestidigitation he was a member of Comptroller Abe Beame's Technical Debt Advisory Committee. His job was to advise on New York's financial transactions. An *expert* should have been more alert to funny stuff. A member of such a committee should have got his back into things and made it his business to find out what was going on. On this a truly truthful man would have confessed to negligence.

But Mr. Simon, if not always rigorous in the pursuit of truth, is adequately uncompromising in the pursuit of principle. He takes full credit for preventing "election-eve political compromises" to ease and improve economic performance in the autumn of 1974, and speaks with near satisfaction of the ensuing loss of forty-three House and three Senate seats by the Republicans. It was in a good cause. Again, in 1976, he has warm praise for the way Gerald Ford followed his advice and that of Alan Greenspan to resist concessions to the electorate right down to election day, although he is more reserved in taking credit for Ford's defeat. He does emerge as a devout follower of Lenin in his belief that a small disciplined body of true believers is far to be preferred to a large, amorphous, compromising majority. There could, one supposes, be a different view by those, such as Mr. Ford, who got sacrificed to such principle.

Mr. Simon extends this commitment to principle to his old friends in Wall Street, and they take a terrible beating. Felix

Rohatyn becomes "Felix-the-Fixer"; David Rockefeller of Chase
Manhattan, Robert Swinarton of Dean Witter and William
Grant of Smith Barney are all "gutless financiers"; and the
worst fate of all befalls Walter Wriston of Citibank and Citicorp. On one page he is recognized "not only as a superb
banker but, also more important, as a financial statesman."
Alas, the short life of the modern financial statesman; a page
later Wriston has "caved in and . . . joined the others," those
others being the gutless crew that wanted a bail-out on New
York.

Mr. Simon could, however, be right in arguing that a New
York default would have been less catastrophic than the bankers
made it out to be and that people who lend money must
expect on occasion to lose it. Also, a point he does not make,
default might have been less serious than some of the cuts in
services and increases in interest charges that were necessary
to avoid it.

He is also right in assigning political power and importance
to the New Class — to the teacher, preacher, writer, scientist,
technocrat and like intellectual. It does, indeed, exercise great
influence. I'm forced to agree here, for Mr. Simon, perhaps
unwittingly, is my disciple. I developed the case in *The Affluent
Society* (in a chapter called "Labor, Leisure and the New Class")
in 1958, having myself taken the phrase from a somewhat different usage by Milovan Djilas. Mr. Simon appears to believe
that the concept comes from Irving Kristol, but Kristol is a
much later convert. I tell all this not out of any abnormal
desire for self-enhancement but to let Mr. Simon know that
he can follow an ex-closet socialist (as he classifies me) and a
uniquely rigorous kind of Communist (as was Djilas) and be
right. Doubtless he will want to try it more often.

Mr. Simon's economic case, to which finally I come, is that
the market has been rejected in favor of bureaucratic planning
and regulation. He attributes this to stupidity, cupidity, weakness of character and bad education by the New Class. He does

not spare his friends in the business world. Here is what he says of them:

> During my tenure at Treasury I watched with incredulity as businessmen ran to the government in every crisis, whining for handouts or protection from the very competition that has made this system so productive. I saw Texas ranchers, hit by drought, demanding government-guaranteed loans; giant milk cooperatives lobbying for higher price supports; major airlines fighting deregulation to preserve their monopoly status; giant companies like Lockheed seeking federal assistance to rescue them from sheer inefficiency; bankers, like David Rockefeller, demanding government bailouts to protect them from their ill-conceived investments; network executives, like William Paley of CBS, fighting to preserve regulatory restrictions and to block the emergence of competitive cable and pay TV. And always, such gentlemen proclaimed their devotion to free enterprise and their opposition to the arbitrary intervention into our economic life by the state. Except, of course, for their own case, which was always unique and which was justified by their immense concern for the public interest.

I find myself in agreement with this; my economic complaint, an objection to some rather overheated writing apart, concerns instead Mr. Simon's further cause and his cure. There is, in fact, a general retreat from the market in our time. The modern large corporation is the outstanding manifestation. So are unions, farmers and their support prices, recipients of the minimum wage, the OPEC countries, the airlines, railroads and truckers. All are part of the escape. This leads, in turn, to various kinds of planned adjustment of supply to demand and a lessening of the sovereignty of the market. It was the rise of OPEC, as Mr. Simon himself agrees, that made him energy czar.

However, Mr. Simon attributes the fall of the market not to this great historical thrust but, as I've noted, to stupidity and related mental or moral aberration or incapacity. Accordingly, he believes that it can be reversed, and the market can be restored by education. Putting his beliefs into practice, he tells proudly in this book that, while in office, he "logged tens of thousands of miles speaking from Miami to Portland, spelling out the danger." For Simon that was pleasant, a source of much agreeable applause, an escape from official drudgery. But it was a terrible waste of time and public money. That applause, as always, came from the already persuaded. As an experienced educator and diligent evangelist, I have learned that economic education is not that powerful, history not that easily reversed. Liberal or conservative, one must come to terms with it. The Secretary should have stayed in Washington and tried his best to make things work. That, not popular spellbinding, is also what Secretaries of the Treasury, Republican or Democratic, are paid for. As this is written, the Democrats are in power. Mr. Simon surely wouldn't want Michael Blumenthal neglecting his work, going up and down the country singing the praises of the mixed economy, the welfare state or, God forbid, anything that could be called closet socialism!

9

Defenders of the Faith, II:
Irving Kristol

I'VE OFTEN THOUGHT how pleasant, easy and also remunerative it would be, were one so motivated, to make the case for modern business organization. One would avoid, above all, the cataleptic and self-refuting litany of the neoclassical market. No one of sound mind, unless extensively conditioned by economic instruction, can any longer be persuaded that the great business firms that constitute the characteristic sector of the modern economy conform to the market-controlled and politically passive model of the neoclassical creed. They have, a dreary and oft-repeated point, that highly visible ability to raise prices. Since watching television is still common, the effort to cross the market and manage the consumer is also widely seen. And papers speak of corporate pressures on the Congress, as does Jimmy Carter, a man who seems safely above any suspicion of being anti-business. The unique foreign policies of the oil companies, ITT and Lockheed have also been well celebrated. So also corporate demands on the United States government. In 1977, the steel companies converted in a matter of weeks from warnings about excessive regulation and big government to demands for more of both to protect them from Japanese competition. The close and symbiotic relationship between the Pentagon and the weapons firms was highlighted nearly twenty years ago by Dwight D. Eisenhower. And, with all else, General Motors, Exxon and IBM simply do not look like the textbook microcosms. So the defense would begin by no longer trying to get people to believe the patently unbelievable about the subordination of the modern large firm to the market and the state, an effort that principally persuades people that there must be something vaguely illegitimate or fraudulent about the large corporation since it tries so elaborately to misrepresent itself.

Having accepted that the corporation transcends its market and has power in the state, then an effective and adequately insouciant defender would say, "What a good thing!" By controlling its prices and managing its customers, it can plan. If hundreds of millions of dollars and several years are to be spent on producing a new automobile model, there must be some assurance as to the eventual price of — and consumer response to — the particular compromise between novelty and banality that is so expensively contrived. If billions are to be put into an oil pipeline across Alaska or a gas pipeline across Canada, there must also be a tight grip on its eventual price. It would be insane to leave that to the unpredictable gyrations of the competitive market. The big corporation eliminates or subdues market forces and substitutes planning. In the modern economy planning is essential.

It would also be held that only a big corporation can deal effectively with big government; the small competitive entrepreneur doesn't have much access even to Michael Blumenthal, James Schlesinger and the other Carter populists, and it is not concern for the small man's state of confidence that has made modern economic administration a minor branch of psychiatry.

Further, as the corporation develops, it takes power away from owners or capitalists and lodges it firmly and irrevocably with the management — the corporate bureaucracy. Capitalists, no one needs to be reminded, were a socially indigestible force

— individualistic, uncompromising, power-hungry, often rapacious, always ready for a fight. Modern corporate bureaucrats, in contrast, are faceless, cautious, courteous, predictable and given to compromise. As a consequence, and because of their power over prices, which they can raise to offset cost increases, we have had in the big-business sector of the economy an unprecedented period of labor peace, though, to be sure, at the price of persistent inflation.

For the same reasons, references to the class struggle, the prelude to the final conflict, have come to have a slightly archaic sound, and among full-time union employees of the thousand largest corporations in the United States there aren't many, perhaps not any, who fall below the poverty line. Nor are there any complaints of shortages of automobiles, depilatories, weapons and other mass products of the corporate sector. This sector extends its technology to agriculture, and the government there follows the corporate practice of giving producers reasonably firm prices against which they can invest and plan. Housing and health care please almost no one, but this could be because they do not lend themselves to large-scale organization and planning.

Large corporations do pursue their own goals and not those of the public in general. And there is the problem of inflation. But the small numbers involved make solution of these problems administratively feasible. In the defense we are now outlining, it would be readily conceded that much must be worked out with the government. And more in the future. Prices and incomes must be restrained if inflation is to be within tolerable limits. And, as the market is replaced by planning, there is no way by which supply is surely adjusted to demand. Energy, where the need for such planning is now accepted in principle if not in actual legislation, is a portent. It is agreed that, with the passage of time, there will have to be an increasing number of planned adjustments of demand to supply. Corporate and public bureaucracy will become increasingly intermingled, as already is the case with oil, atomic energy and space adventure. The corporation, it would be said, paves the way for the requisite national planning and, by removing power from the owners, even goes far toward socializing itself.

Finally, it would be part of this defense that while in the manner of all disciplined organizations the corporation denies liberty of expression to quite a few people, those so oppressed do not in the least seem to mind. The repression operates against the important participants in the management. In all public utterances they are well advised to reflect an institutionally acceptable viewpoint, what is often called, even in non-Communist countries, the party line. Very often they are required to say what the guardians of the official truth, the public relations executives, have written out for them — a derogatory thing. And the sanctions for speaking in conflict with the corporate perception of truth, if subtle, are severe: those so disposed may be praised for their character, but they are not promoted. That, over time, becomes expensive.

There is also, in the upper ranks of the corporation, an equally subtle discouragement of other forms of expression. Here is an exceptionally privileged and compensated group of men which produces almost no poets, essayists, novelists, painters, philosophers or composers and, as compared with the law and the universities, very few politicians. The understanding, never explicit, is that such expression is to be avoided. A man whose interests and energies are partly so engaged may again be praised, but he is also thought a trifle eccentric and unreliable. So it is part of the corporate discipline that he set such matters aside. The repression appears almost innocently in the dictum: "We expect a man around here to give his full time to his job."

But it can be held on behalf of modern enterprise that this repression, however real, is loved. The pecuniary rewards are large enough to compensate for the surrender of individual views to corporate thought and expression. The pecuniary and social costs of contracting out are high but less painful than in the formally planned economies. And there is even, as John Steinbeck once said I would find of the State Department, a measure of contentment and comfort in knowing that thought is available, along with all else, from the organization.

Meanwhile thought and expression by people outside the corporation are not controlled or condemned. They may be considered inconvenient or annoying, but the modern corporate man, in contrast with the old-fashioned capitalist entrepreneur, does nothing about it. A generation ago every reputable university board had a few vintage entrepreneurs who were stoutly in search of heresy in the economics department. The modern corporate man would not dream of intervening, even though, as I shall note presently, there is some thought that, financially speaking, he should.

All in all, it's quite a good case, and it only becomes possible when one has dispensed with the conventionally implausible authority of the market. Much is missing, including such matters as distribution of income, distribution of power, the position of minorities, social values and purposes, the tendency of the well-articulated desires of corporate executives and the generally privileged to be mistaken for the needs of the masses, and the risks that are taken of nuclear extinction. But I have been concocting a sample brief, not offering a balanced view.

Were I asking for someone to make this case, I've long thought it would be Irving Kristol. He is Henry Robinson Luce Professor of Urban Values at New York University, and his academic auspices[1] are exactly right. Harry Luce, in a groping, erratic, charming way, spent much of his adult life

---

[1] A powerful influence, this. For much of my life I was Paul M. Warburg Professor of Economics at Harvard. The donor of the chair, the late James Warburg, was, by most calculations, a socialist. Mr. Kristol in his writings is almost tearfully distressed that I am not more generally recognized to be such.

looking for a rationale for the kind of big businessman that, to his unending astonishment, he had become. Mr. Kristol is also far too sophisticated to buy the primitive neoclassical and free enterprise model in its Simple or William Simon form. He has even the good judgment to put himself at some distance from Milton Friedman. A further qualification, which honesty requires me to record, is the breadth of his reading, which, as *Two Cheers for Capitalism*[2] makes abundantly, even repetitively clear, extends to all of my books, and with nearly total recall. Only a greater measure of agreement might be wished.

Alas, although Kristol tried, this recent book falls short of making the case, as I trust even less partisan judges will find to be so. Mr. Kristol accepts the decline of the market and agrees that the modern corporation has political power. And he is intensely critical of the way both power over prices and power in political matters are used or not used. Thus he thinks the oil companies, after the Arab boycott and the OPEC price hike, were insane to rake in the money while explaining that profits were not as high as they once were or should be. It is, obviously, their decision he is attacking, not the impersonal award of the market. He also believes corporate executives are more interested in excluding their stockholders from power in the corporation than in mobilizing them as a constituency for political ends, and he sees much other misused or unused political power. He concedes that the managerial revolution is real and that the corporate technostructure *is* the decisive political force. One could not ask for more on these matters.

But having gone all this distance, he then retreats, I would say abandons his real defenses, and makes the neoclassical market the bulwark of his defense of capitalism. He notes that "Americans who defend the capitalist system, i.e., an economy and a way of life organized primarily around the free market, are called 'conservative,' " and he leaves no doubt that, subject

2 (New York: Basic Books, 1978.)

to some "neo-conservative" concessions to government intervention, this is how he classifies himself. It is the peculiar national sin of American liberals, he asserts, that they do not see that the market is on their side. And even more powerfully he avows that "One of the *keystones* [my emphasis] of modern economic thought is that it is impossible to have an *a priori* knowledge of what constitutes happiness for other people; that such knowledge is incorporated in an individual's 'utility schedules'; and this knowledge, in turn, is revealed by the choices the individual makes in a free market."

This particular defense is truly unwise. For if, as Mr. Kristol concedes, the corporation can influence prices, those absolutely indispensable "utility schedules" operate not against the free market that he so specifically requires but, in practice, against a rigged price schedule, one that reflects the economic power of the corporation. In consequence, the corporation deeply influences consumer choice through the relative prices it charges. (This result is wholly conceded even in neoconservative, neoclassical economic thought and is readily verified if one thinks of the effect that decisions on oil prices, for example, have on consumer choice.) And if, as Mr. Kristol partly agrees — and would have difficulty wholly denying — the modern firm spends hugely to influence consumer taste, that sacrosanct utility schedule, already destroyed by corporate price-fixing, is itself partly the creation of the corporation.

So Mr. Kristol is a truly devastating force against his own keystones. He could have argued that, in a relatively rich society, many of the prices and tastes so influenced by the corporation have no very profound effect on well-being — that the consumer is malleable because, being well supplied, his or her needs are not so urgent as to command deep thought. But this he does not stress, and there are limits beyond which, even in the most gracious of moods, one cannot go in making another author's case.

In Mr. Kristol's case for capitalism the corporation is the

enemy of the market by which he then defends the corporation. But this curious result, not surprisingly, he does not wish to concede, perhaps even to himself. So, since he cannot rescue the corporation by restoring the market, he uncovers another inimical force from which it needs to be saved. This is professors, journalists, scientists and assorted pundits — a community which, accepting my earlier lead,[3] for which I am duly grateful, he groups together as the New Class. The New Class reacts adversely to the corporation partly out of a sense of inferiority, partly out of envy of executive pay and expense accounts and partly out of a defensive reaction to the economic achievements of the corporation and specifically to its service to the consumer economy. This last, in contrast with the concerns of the mind or soul, the New Class holds to be vulgar. Derived from these attitudes are an excessively costly concern for the environment (which Mr. Kristol, after conceding the adverse effect, would leave entirely to his imperfectly functioning market), a prodigal attitude toward public services and the welfare state and a generally feckless commitment to closing tax loopholes and redistributing income. The need, he believes, is for a strenuous effort by business and its friends to educate the New Class, although he has only contempt for past business efforts to promote a better understanding of free enterprise, and, characteristically, he thinks advertisements and other such efforts do as much harm as good.

The nature of the desirable remedial education, which one imagines will be a fairly challenging task, Mr. Kristol does not make completely clear. He does say that corporations should distinguish between their friends and their critics in distributing corporate largesse. And, as to both this and the remedial education, he urges that they be guided by their friends within the New Class, a convocation that almost certainly would include Mr. Kristol.

[3] See the earlier essay, "Defenders of the Faith, I: William Simon."

One wonders, however, if such discrimination would not stir a certain amount of adverse comment and even antagonism in the New Class, although I do agree with Mr. Kristol that a fair number of our Classmates are open to purchase if the price is sufficiently high. I would be more troubled by his diagnosis. Over the years, as a certified member, I have not noticed the envy or sense of inferiority of which he speaks. My impression, rather, is of people who tend, if at all, to a definite self-regard and self-assurance. Maybe Mr. Kristol, during the course of his researches, should have made a brief field trip to the Harvard Faculty Club in Cambridge. I would judge that our Classmates manifest a concern for corporate behavior, as for government, not out of envy but because they sense rightly that this is where the power lies. Were trade unions, the Baptist Church or *The Public Interest* (edited by Mr. Kristol) equally powerful, they would be equally the object of attention.

But the ultimate question in educating the New Class gets back to what case can be sold. Here Mr. Kristol has his greatest reason to worry. He is too sophisticated and open to the evidence to accept the case for the market. But being a conservative in mood as well as in politics, he cannot bring himself to sell anything else. So, having destroyed the case for the market, he goes back to it.

This is generally thought to be a cold season for liberals. Mr. Kristol, in a literate and learned way, has shown that it is also a hard time for conservatives.

10

Defenders of the Faith, III:
Wright and Slick

SOME TIME AGO a small volume — nine pages in all — came
to me in the mail. Its author was Mr. M. A. Wright, then
Chairman of Exxon, U.S.A. The title, arresting and even alarming, was *The Assault on Private Enterprise*. My first thought
was that another business executive was trying to communicate.
On the first page was the telltale sentence: "Business has failed
to do an effective job in communicating its point of view to the
general public." All disasters in executive prose begin with
these words. Then came the compulsive cry: "Let there be no
mistake: an attack is being mounted on the private enterprise
system in the U.S. *The life of that system is at stake.*" (The
emphasis on the overstatement is mine.) These are the trade-
marks, invariable, utterly reliable, of executive communication.

Politicians have a curious contempt for the art of conveying
thought. A surprising number give speeches over a lifetime
without ever learning to make, write or even read a speech.
Lecture-circuit impresarios are worse; all feel that nothing so
moves an audience (and justifies their fee) as a suffocating re-
cital of the commonplace, combined with a deeply condescending manner. Social scientists, although perhaps improving, have
their own special instinct for fraternal obscurity. But there is
no doubt that business executives are the worst of all. Theirs

is the egregiously optimistic belief that people will believe anything, however improbable, if it is said with emphasis and solemnity by the head of a really big company. With the ultimate corporate promotion comes the right to proclaim truth.

Mr. Wright, I concluded, was a man of orthodox and hence profound, even perverse, inadequacy in communication. But the pamphlet, and the circumstances of its circulation, led me to wonder if something more than normal executive incoherence might be involved. I was sent the document by Mr. W. T. Slick, Jr., a senior vice president of Exxon, U.S.A. He sent it with a transmittal note summarizing Mr. Wright's main point: all business, not just the oil companies, is being bombed by its enemies, and a great closing of ranks is called for. Then, significantly, Mr. Slick went on to hope that readers would find the document "highly thought-provoking." This, perhaps even more than Mr. Slick's name, seemed the clue. No friend of free enterprise could conceivably want to provoke thought on Mr. Wright's dialectics. One could only conclude that the enemies of free enterprise do abound and that one of them, Mr. Slick, was right inside Exxon. The evidence was and remains circumstantial but, nonetheless, overwhelming.

Mr. Wright began by stressing the climate of "suspicion and, it must be said, hostility" in which the business firm now operates. Then he went on to offer the first proposition on which the sinister Mr. Slick was out to provoke thought. It was that in the sixties there was the war in Vietnam, the riots in the cities, then inflation, then pollution or, in Mr. Wright's more awesome words, "a belated recognition of the burden placed on the quality of air and water by our highly industrialized society." All this got people in a foul mood which extended to their attitudes toward business. Then came an outburst of dissatisfaction with consumer products — Americans got discouraged with the quality, design, safety, reliability and dealer-servicing of the things they bought, and, being in this state of mind, they fell prey to consumerism. Finally came Watergate which crystallized public distrust of almost everything and everybody. While all this was happening, according to Mr. Wright, most business executives were innocently going their own way — "too busy," he said, "competing in the private enterprise system to concern themselves with communicating its virtues."

Mr. Slick, I concluded, could not possibly be unaware of the consequences of provoking or otherwise stimulating thought on these propositions. He was a senior vice president, after all. No, he knew that Mr. Wright's plea would surely remind people of those old arguments over who makes money out of wars. More often a corporation than a slogging foot soldier. And it would make them wonder why Exxon didn't consider cities and their people part of the system. And it would get people's minds strongly on the effect of oil prices on inflation. And, where pollution is concerned, on gasoline fumes and oil on the beaches and those big, high oil-company signs on the roadsides. And what guileless oil man would urge thought on that really big one about the tycoons being too busy competing to notice even their own prices?

Slick was being yet more obviously subversive of Wright when he invited reflection on the consumer movement and the ungreening of the consumer. Stupidity is still a problem in this Republic. But neither he nor anyone in his position could expect people to believe that manufacturers are without responsibility for the quality, durability, design and safety of their products, however busy they are competing in the free enterprise system.

Slick came even closer to blowing his cover on Watergate. What kind of vice president would want thought on where the money came from for the burglars? (Most from oil, the rest from soybean oil.) And on who helped to launder it through Mexico? (Oil men again.) And where the biggest single bundle of cash came from for the cover-up? (Northrop this time.) And why would he want to put people's minds on those secret contributions by corporate executives and the illegal political contributions by the corporations themselves?

Mr. Slick got in some other, less transparent blows. Thus, Mr. Wright was indignant about suggestions that the depletion allowance be abolished. No well-intentioned vice president would invite reflection on the free ride thus being given to the oil companies. Mr. Wright was dismayed by suggestions that the foreign tax credits on income earned overseas be disallowed. This provoked an especially dangerous new line of thought, for what the oil companies call a tax credit and subtract from their tax bill is really a royalty payment that should come off their costs. And Mr. Wright cited James Buckley of New York and Paul Fannin of Arizona as the two statesmen who still defended the oil company position at the time. That these were the available advocates is a small but damaging detail on which to have people pondering.

Mr. Wright then went on to warn that government is steadily moving in on private enterprise. In urging thought on that, Mr. Slick must have known that the most retarded or reluctant mind would turn to Lockheed, Penn Central, the eastern railroads and the Franklin National Bank, all, in fact, examples of private enterprise moving strongly in on the government. Mr. Wright inquired if there wouldn't some day be a U.S. Automobile Corporation or a Federal Steel Company. People who reflect will conclude that sure as hell there will be. It will happen on precisely the day when a big enough automobile company or a big enough steel company loses a big enough sum of money so that a rescue operation is demanded of the good old federal government.

Finally came Mr. Slick's real coup — a small miracle, among other things, of timing. Mr. Wright argued that the public has a ridiculously exaggerated view of the earnings of companies such as his, and being economically illiterate (my phrase), they "do not understand the role of profit or the function of the free market." A few days after these words were circulated, Exxon's second-quarter earnings report was published in the paper, and no one can say that Mr. Slick didn't know it was coming. It told that profits (these were for the company as a whole) were up 59 percent over the previous year and that for the first six months they were up 52.8 percent, for the second quarter 66.7 percent, in each case also as compared with the year before. All records had been broken.

With a supporter like Slick, one wonders if either Wright or free enterprise needs enemies.

There is further possible proof of the point, although something, no doubt, can be attributed to natural causes. Not long after the above was written, Wright was superseded. So was Slick.

11

Who Was Thorstein Veblen?

There is always an aura of playfulness about his attitude toward his own work in marked contrast to the deadly seriousness of most economists.

Wesley C. Mitchell

THE NEAREST THING in the United States to an academic legend — the equivalent of that of Scott Fitzgerald in fiction or of the Barrymores in the theater — is the legend of Thorstein Veblen. A legend is reality so enlarged by imagination that, eventually, the image has an existence of its own. This happened to Veblen. He was a man of great and fertile mind and a marvelously resourceful exponent of its product. His life, beginning on the frontier of the upper Middle West in 1857 and continuing, mostly at one university or another, until 1929, was not without romance of a kind. Certainly by the standards of academic life at the time it was nonconformist. There was ample material both in his work and in his life on which to build the legend, and the builders have not failed. There is also a considerable debt to imagination.

What is believed about Veblen's grim, dark boyhood in a poor, immigrant Norwegian family in Wisconsin and Minnesota, his reaction to those oppressive surroundings, his harried life in the American academic world in the closing decades of the last century and the first three of this, the fatal way he attracted women and vice versa and its consequences in his tightly corseted surroundings, the indifference of all right-thinking men to his work, has only a limited foundation in fact.

Economics is a dull enough business, and sociology is sometimes worse. So, on occasion, are those who teach these subjects. One reason they are dull is the belief that everything associated with human personality should be made as mechanical as possible. That is science. Perhaps one should, instead, perpetuate any available myth. When, as with Veblen, the man is enlarged by a nimbus, the latter should be brightened, not dissolved.

Still, there is a certain case for truth, and the facts in Veblen's case are also far from tedious. He is not, as some have suggested, a universal source of insight on American society. Like Smith and Marx, he did not see what had not yet happened. Also, on some things, he was wrong, and faced with a choice between strict accuracy and what would outrage his audience, he rarely hesitated. But no man of his time, or since, looked with such a cool and penetrating eye at pecuniary gain and the way its pursuit makes men and women behave.

This cool and penetrating view is the substance behind the Veblen legend. It is a view that still astonishes the reader with what it reveals. While there may be other deserving candidates, only two books by American economists of the nineteenth century are still read. One of these is Henry George's *Progress and Poverty*; the other is Veblen's *The Theory of the Leisure Class*. Neither of these books, it is interesting to note, came from the sophisticated and derivative world of the eastern seaboard. Both were the candid, clear-headed, untimid reactions of the frontiersman — in the case of Henry George to speculative alienation of land, in the case of Veblen to the pompous social ordinances of the affluent. But the comparison cannot be carried too far. Henry George was the exponent of a notably compelling idea; his book remains important for that idea — for the notion of the high price that society pays for private ownership and the pursuit of capital gains from land. Veblen's great work is a wide-ranging and timeless comment on the behavior of people who possess or are in pursuit of wealth and who, looking beyond their possessions, want the eminence that, or so they believe, wealth was meant to buy. No one has really read very much social science if he hasn't read *The Theory of the Leisure Class* at least once. Not many of more than minimal education and pretense get through life without adverting at some time or other to "conspicuous consumption," "pecuniary emulation" or "conspicuous waste," even though they may not know whence these phrases came.

Veblen's parents, Thomas Anderson and Kari Bunde Veblen, emigrated from Norway to a farm in rural Wisconsin in 1847, ten years before Thorstein's birth. There were the usual problems in raising the money for the passage, the quite terrible hardships on the voyage. In all, the Veblens had twelve children, of whom Thorstein was the sixth. The first farm in Wisconsin was inferior to what later and better information revealed to be available farther west. They moved, and, in 1865, they moved a second time. The new and final holding was on the prairie, now about an hour's drive south from Minneapolis. It is to this farm that the legend of Veblen's dark and deprived boyhood belongs. No one who visits this countryside will believe it. There can be no farming country in the world with a more generous aspect of opulence. The prairie is gently rolling. The soil is black and deep, the barns are huge, the silos numerous and the houses big, square and comfortable, without architectural ambition. The Veblen house, with a long view of the surrounding farmland, is an ample, pleasant, white frame structure bespeaking not merely comfort but prosperity.[1]

---

[1] In recent decades it had fallen on difficult times. Partly as the result of a television program based on this essay, and my plea to then Governor Wendell Anderson — "only Scandinavians are so negligent of their heroes" — it was acquired as a national historic site, and rehabilitation is in prospect.

Families in the modern suburban tract are not housed as well.

Since this countryside was originally open, well-vegetated prairie, it must have looked very promising to the settler a hundred years ago. Thomas Veblen acquired 290 acres of this wealth; it is hard to imagine that he, his wife or, by their instruction, any of their children could have thought of themselves as poor. Not a thousand, perhaps not even a hundred farmers working their own land were so handsomely provided in the Norway they had left. Nor, in fact, did the Veblens think themselves poor. Later, in letters, Thorstein's brothers and sisters were to comment, sometimes with amusement, on occasion with anger, on the myth of their poverty-stricken origins.[2]

If this part of the Veblen story is unremarkable and commonplace — the tearing up of roots, departure, hardship, miscalculation, eventual reward — there were other things that separated the Veblens from the general run of Scandinavian settlers and that help to explain Thorstein. Thomas Veblen, who had been a skilled carpenter and cabinetmaker, soon proved himself a much more than normally intelligent and progressive farmer. And, however he viewed the farm for himself, he almost certainly regarded it as a steppingstone for his children. Even more exceptional was Kari, his wife. She was an alert, imaginative, self-confident and intelligent woman, lovely in appearance, who, from an early age, identified, protected and encouraged the family genius. In later years, in a family and community where more hands were always needed and virtue was associated, accordingly, with efficient toil — effectiveness as a worker was what distinguished a *good* boy or girl from the rest — Thorstein Veblen was treated with tolerance. Under the cover of a weak constitution he was given leisure for reading. This released time could only have been provided by remarkably perceptive parents. One of Veblen's brothers

[2] The letters are in the archives of the Minnesota Historical Society in St. Paul, to which I am grateful.

later wrote that it was from his mother that "Thorstein got his personality and brains," although others thought them decidedly his own property.

Thorstein, like his brothers and sisters, went to the local schools, and, on finishing these, to Carleton College (then styled Carleton College Academy) in the nearby town of Northfield. His sister Emily was in attendance at the same time; other members of the family also went to Carleton. In an engaging and characteristic move their father acted to keep down college expenses. He bought a plot of land on the edge of town for the nominal amount charged for such real estate in that time and put up a house to shelter his offspring while they were being educated. The Veblen legend further holds that the winning of an education involved Thorstein in major and even heroic hardship. This can be laid decisively to rest. A letter in the archives of the Minnesota Historical Society from Andrew Veblen, Thorstein's brother, notes that money was available, if not abundant: "Father gave him the strictly necessary assistance through his schooling. Thorstein, like the rest of the family, kept his expenses down to the minimum. . . all in line with the close economy that the whole family practiced."[3] A sister-in-law, Florence (Mrs. Orson) Veblen, wrote more indignantly and with a characteristic view of what was virtue in those times: "There is not the slightest ground for depriving my father-in-law of the credit of having paid for the education of his children — *all* of them — he was well able to do so; he had two good farms in the richest farming district in America."[3]

It was, nevertheless, an exception to the general community practice that the Veblen children should be sent to college rather than be put to useful work, as Norwegian farmers would

[3] These letters were written in 1926 to Joseph Dorfman, Veblen's distinguished biographer. Dorfman, alas, does something to perpetuate the legend of deprivation.

then have called it, on the farm. Exceptionally too they were sent to an Anglo-Saxon denominational college — Carleton was Congregationalist — rather than to one of the Lutheran institutions which responded to the language, culture and religion of the Scandinavians. The Veblen myth (as the family has insisted) has also exaggerated somewhat the alienation of the Norwegians in general and the Veblens in particular in an English-speaking society. It is part of the legend that Thorstein's father spoke no English and that his son had difficulty with the language. This is nonsense. But in the local class structure the Anglo-Saxons were the dominant town and merchant class, the Scandinavians the hard-working peasantry. The Veblen children were not intended for their class.

Carleton was one of the denominational colleges which were established as the frontier moved westward and by which it was shown that along with economic and civic achievement in America went also culture and religion. Like the others of its age, it was unquestionably fairly bad. But, like so many small liberal arts colleges of the time, it was the haven for a few learned men and devoted teachers — the saving remnant that seemed always to show up when one was established. Such a teacher at Carleton in Veblen's time was John Bates Clark, later, when at Columbia University, to be recognized as the dean of American economists of his time. (He was one of the originators of the concept of marginality — the notion that decisions concerning consumption are made not in consequence of the total stock of goods possessed but in consequence of the satisfactions to be derived from the possession or use of another unit added to the stock already possessed.) Veblen became a student of Clark's; Clark thought well of Veblen.

This must have required both imagination and tolerance, for in various of his class exercises Veblen was already giving indication of his later style and method. He prepared a solemn and ostentatiously sincere classification of men according to their noses; one of his exercises in public rhetoric defended the drunkard's view of his own likely death; another argued the case for cannibalism. Clark, who was presiding when Veblen appeared to favor intoxication, felt obliged to demur. In a denominational college in the Midwest at the time, cannibalism had a somewhat higher canonical sanction than intoxication. Veblen resorted to the defense that he was to employ with the utmost consistency for the rest of his life. No value judgment was involved; he was not being partial to the drunk; his argument was purely scientific.

Veblen finished his last two years at the college in one and graduated brilliantly. His graduation oration was entitled "Mill's Examination of Hamilton's Philosophy of the Conditioned." It was described by contemporaries as a triumph, but it does not survive. While at Carleton Veblen had formed a close friendship with Ellen Rolfe; she was the daughter of a prominent and prosperous Midwestern family and, like Veblen, was independent and introspective, very much apart from the crowd and also highly intelligent. They were not married for another eleven years, although this absence of haste did not mean that either had any less reason to regret it in later leisure. The legend holds Veblen to have been an indifferent and unfaithful husband who was singularly incapable of resisting the advances of the women whom, however improbably, he attracted. The Veblen family seems to have considered the fault to be at least partly Ellen's. She had a nervous breakdown following an effort at teaching; in a far from reticent and, one supposes, deeply partisan letter in the St. Paul archives,[4] a sister-in-law concludes: "There is not the least doubt that she

---

4 From Florence Veblen, 1926. In an earlier account of the Veblen family (Orson Veblen, "Thorstein Veblen: Reminiscences of His Brother Orson," *Social Forces*, vol. X, no. 2 [December 1931], pp. 187–195), Florence Veblen also dealt harshly with Ellen. However, an unpublished comment (again in the St. Paul archives) by Andrew Veblen dissents at least in part, noting also that the two women had never met.

is insane." Only one thing is certain: it was an unsuccessful and unhappy marriage.

After teaching for a year at a local academy following his graduation from Carleton, Veblen departed for Johns Hopkins University in Baltimore to study philosophy. At this time, 1881, Johns Hopkins was being advertised as the first American university with a specialized postgraduate school following the earlier European design. The billing, as Veblen was later to point out, was considerably in advance of the fact. Money and hence professors were very scarce; the Baltimore context was that of a conservative southern town. Veblen was unhappy and did not complete the term. He began what — with one major interruption — was to be a lifetime of wandering over the American academic landscape.

His next stop was Yale. It was an interesting time in New Haven — what scholars inclined to metaphors from the brewing industry call a period of intellectual ferment. One focus of contention was between Noah Porter, a seemingly pretentious divine then believed to be an outstanding philosopher and metaphysician, and William Graham Sumner, the American exponent of Herbert Spencer. The practical thrust of Noah Porter's effort was to prevent Sumner from assigning Spencer's *Principles of Sociology* to his classes. In this he prevailed; Spencer was righteously suppressed. Porter's success, one imagines, proceeded less from the force of his argument against Spencer's acceptance of the Darwinian thesis — natural selection, survival of the fittest — as a social and economic axiom than from the unfortunate fact that he was also the president of the university. In Veblen's later writing there is a strong suggestion of Spencer and Sumner. Natural selection is not the foundation of Veblen's system, but it serves him as an infinitely handy explanation of how some survive and prosper and others do not.

There has been much solemn discussion of the effect on Veblen's later writing of the philosophical discussion at Yale and of his own dissertation on Kant. My instinct is to think it was not too great. This is affirmed in a general way by the other Veblens. In later years his brother Andrew (a physicist and mathematician) responded repeatedly and stubbornly, though, no doubt with some exaggeration, to efforts to identify the sources of Thorstein Veblen's thought: "I do not think that any person much influenced the formation of his views or opinions."[5]

After two and a half years at Yale — underwritten by a brother and the Minnesota family and farm — Veblen emerged with a Ph.D. He wanted to teach; he also had rather favorable recommendations. But he could not find a job and so he went back to the Minnesota homestead. There, endlessly reading and doing occasional writing, he remained for seven years. As in his childhood, he again professed ill health. Andrew Veblen, later letters show, thought the illness genuine; other members of his family diagnosed his ailment as an allergy to manual toil. He married, and Ellen brought with her a little money. From time to time he was asked to apply for teaching positions; tentative offers were righteously withdrawn when it was discovered that he was not a subscribing Christian. In 1891, he resumed his academic wandering by registering as a graduate student at Cornell.

The senior professor of economics at Cornell at the time was J. Laurence Laughlin, a stalwart exponent of the English classical school, who, until then, had declined to become a member of the American Economic Association in the belief that it had socialist inclinations. (There has been no such suspicion in recent times.) Joseph Dorfman of Columbia University, an eminent student of American economic thought and the preeminent authority on Veblen, tells of Laughlin's meeting with Veblen in *Thorstein Veblen and His America,* a massive book

[5] Letter in the St. Paul archives.

to which everyone who speaks or writes on Veblen is indebted.[6,7] Laughlin "was sitting in his study in Ithaca when an
anemic-looking person, wearing a coonskin cap and corduroy
trousers, entered and in the mildest possible tone announced: 'I
am Thorstein Veblen.' He told Laughlin of his academic history, his enforced idleness, and his desire to go on with his
studies. The fellowships had all been filled, but Laughlin was
so impressed with the quality of the man that he went to the
president and other powers of the university and secured a
special grant."[8]

Apart from the impression that Veblen's manner and dress so
conveyed, the account is important for another reason. Always
in Veblen's life there were individuals — a small but vital
few — who strongly sensed his talents. Often, as in the case of
Laughlin, they were conservatives — men who, in ideas and
habits of life, were a world apart from Veblen. Repeatedly
these good men rescued or protected their prodigious but always inconvenient friend.

Veblen was at Cornell rather less than two years, although
long enough to begin advancing his career with uncharacteris-
tic orthodoxy by getting articles into the scholarly journals.
Then Laughlin was invited to be head of the Department of
Economics at the new University of Chicago. He took Veblen
with him; Veblen was awarded a fellowship of $520 a year for
which he was to prepare a course on the history of socialism
and assist in editing the newly founded *Journal of Political*

6 Although members of the family have disputed Dorfman on numerous details.
In the library of the Minnesota Historical Society there is a heavily annotated
copy of Dorfman's book giving Emily Veblen's corrections and dissents. Numerous
minor points of family history are challenged; like other members of the family,
she protested all suggestions that the family was poor or that it was alienated
from the rest of the community. And she thought that Dorfman's picture of
Thorstein as a lonely, shabby, excessively introverted boy and student was much
overdrawn.

7 Joseph Dorfman, *Thorstein Veblen and His America* (New York: Viking, 1934).
8 Dorfman, pp. 79–80.

Economy. He was now thirty-five years old. In the next several years he advanced to the rank of tutor and instructor, continued to teach and to edit the *Journal* (known to economists as the J.P.E.), wrote a great many reviews and, among other articles, one on the theory of women's dress, another on the barbarian status of women and a third on the instinct of workmanship and the irksomeness of labor. All these foreshadowed later books.

In these years he also developed his teaching style, if such it could be called. He sat at a table and spoke in a low monotone to the handful of students who were interested and who could get close enough to hear. He also discovered, if he had not previously learned, that something — mind, manner, dress, his sardonic and challenging indifference to approval or disapproval — made him extremely attractive to women. His wife now found that she had more and more competition for his attention. It was something to which neither she nor the academic communities in which Veblen resided ever reconciled themselves.

In 1899, while still at Chicago and while Laughlin was still having trouble getting him small increases in pay or even, on occasion, getting his appointment renewed, he published his first and greatest book. It was *The Theory of the Leisure Class.*

There is little that anyone can be told about *The Theory of the Leisure Class* that he or she cannot learn better by reading the book himself or herself. It is a marvelous thing and, in its particular way, a masterpiece of English prose. But the qualification is important. Veblen's writing cannot be read like that of any other author. Wesley C. Mitchell, regarded, though not with entire accuracy, as Veblen's leading intellectual legatee, once said that "One must be highly sophisticated to enjoy his [Veblen's] books." [9] Those who cherish Veblen would like, I

---

[9] Wesley C. Mitchell, *What Veblen Taught* (New York: Viking, 1936), p. xx.

am sure, to believe this. The truth is of a simpler sort. One needs only to realize that, if Veblen is to be enjoyed, he must be read very carefully and very slowly. He enlightens, amuses and delights but only if he is given a good deal of time.

That is because one cannot divorce Veblen's ideas from the language in which they are conveyed. The ideas are pungent, incisive and insulting. But the writing is a weapon as well. Veblen, as Mitchell also noted, writes "with one eye on the scientific merits of his analysis, and his other eye fixed on the squirming reader."[10] And he startles his reader with an exceedingly perverse use of meaning. This never varies from that sanctioned by the most precise and demanding usage, but in the context it is often unexpected. His usage Veblen then attributes to scientific necessity. Thus, in his immortal discussion of conspicuous consumption, he notes that expenditure, if it is to contribute efficiently to the individual's "good fame," must generally be on "superfluities." "In order to be reputable it [the expenditure] must be wasteful."[11] All of this is quite exact. The rich do want fame; reputable expenditure is what adds to their repute or fame; the dress, housing, equipage, that serve this purpose and are not essential for existence are superfluous. Nonessential expenditure is wasteful. But only Veblen would have used the words "fame," "reputable" and "waste" in such a way. In the case of "waste," he does decide that some explanation is necessary. This is characteristically airy and matter-of-fact. In everyday speech, "the word carries an undertone of deprecation. It is here used for want of a better term ... and it is not to be taken in an odious sense ..."[12]

And in a similar vein: The wives of the rich forswear useful employment because "Abstention from labor is not only an honorific or meritorious act, but it presently comes to be a

---

[10] Mitchell, p. xviii.

[11] Thorstein Veblen, *The Theory of the Leisure Class* (Boston: Houghton Mifflin, 1973), p. 77.

[12] Thorstein Veblen, p. 78.

requisite of decency."[13] "Honor," "merit" and "decency" are all used with exactness, but they are not often associated with idleness. A robber baron, Veblen says, has a better chance of escaping the law than a small crook because "A well-bred expenditure of his booty especially appeals . . . to persons of a cultivated sense of the proprieties, and goes far to mitigate the sense of moral turpitude with which his dereliction is viewed by them."[14] Scholars do not ordinarily associate the disposal of ill-gotten wealth with good breeding.

One sees also from this sentence why Veblen must be read slowly and carefully. If one goes rapidly, words will be given their usual contextual meaning — not the precise and perverse sense that Veblen intended. Waste will be wicked and not a source of esteem; the association of idleness with merit, honor and decency will somehow be missed, as well as that of the social position of the crook with the public attitude toward his expenditure. *The Theory of the Leisure Class* yields its meaning, and therewith its full enjoyment, only to those who also have leisure.

When Veblen had finished his manuscript, he sent it to the publisher, and it came back several times for revision. Eventually, it appears, publication required a guarantee from the author. The book could not have been badly written in any technical or grammatical sense. Veblen by then was an experienced editor. Nor was he any novice as a writer. One imagines that the perverse and startling use of words, combined no doubt with the irony and the attack on the icons, was more than the publisher could readily manage. But, on the other side, some very good reader or editor must have seen how much was there.

The Theory of the Leisure Class is a tract, the most comprehensive ever written, on snobbery and social pretense. Some

13 Thorstein Veblen, p. 44.
14 Thorstein Veblen, p. 89.

of it has application only to American society at the end of the
last century — at the height of the gilded age of American
capitalism. More is wonderfully relevant to modern affluence.

The rich have often been attacked by the less rich because
they have a superior social position that is based on assets and
not on moral or intellectual worth. And they have also been
accused of using their wealth and position to sustain a profligate
consumption of resources of which others are in greater need,
and of defending the social structure that accords them their
privileged status. And they have been attacked for the base,
wicked or immoral behavior that wealth sustains and their
social position sanctions. These attacks the rich can endure.
That is because the assailants concede them their superior
power and position; they only deny their right to that position
or to behave as they do therein. The denial involves a good
deal of righteous anger and indignation. The rich are only
reminded that they are thought worth such anger and indignation.

Here is Veblen's supreme literary and polemical achievement. He concedes the rich and the well-to-do nothing; and
he would not dream of suggesting that his personal attitudes or
passion are in any way involved. The rich are anthropological
specimens; the possession of money and property has made
their behavior interesting and visibly ridiculous. The effort to
establish one's importance and precedence and the yearning for
the resulting esteem and applause are matters only of sociologi-
cal and anthropological interest and are common to all humans. Nothing in the basic tendency differentiates a Whitney,
Vanderbilt or Astor from a Papuan chieftain or "for instance,
the tribes of the Andamans." The dress, festivals or rituals and
artifacts of the Vanderbilts are more complex; the motivation
is in no way different from or superior to that of their barbarian counterparts.

That is why the rich are not viewed with indignation. The
scientist does not become angry with the primitive tribesman
because of the extravagance of his sexual orgies or the venge-

ance of his self-mutilation. So with the social observances of the rich. Their banquets and other entertainments are equated in commonplace fashion with the orgies; the self-mutilation of the savage is the equivalent of the painfully constricting dress in which, at that time, the women of the well-to-do were corseted.

One must remember that Veblen wrote in the last years of the last century — before the established order suffered the disintegrating onslaught of World War I, V. I. Lenin and the leveling oratory of modern democratic politics. It was a time when gentlemen still believed they were gentlemen and that it was mostly wealth that made the difference. Veblen calmly identified the manners and behavior of these so-called gentlemen with the manners and behavior of the people of the bush. Speaking of the utility of different observances for the purpose of affirming or enhancing the individual's repute, Veblen notes that "Presents and feasts had probably another origin than that of naive ostentation, but they acquired their utility for this purpose very early, and they have retained that character to the present . . . Costly entertainments, such as *the potlatch or the ball*, are peculiarly adapted to serve this end."[15] The italics equating the potlatch and the ball are mine; Veblen would never have dreamed of emphasizing so obvious a point.

While *The Theory of the Leisure Class* is a devastating put-down of the rich, it is also more than that. It brilliantly and truthfully illuminates the effect of wealth on behavior. No one who has read this book ever again sees the consumption of goods in the same light. Above a certain level of affluence the enjoyment of goods — of dress, houses, automobiles, entertainment — can never again be thought intrinsic as, in a naive way, the established or neoclassical economics still holds it to be. Possession and consumption are the banner that advertises achievement, that proclaims, by the accepted standards of the

15 Thorstein Veblen, p. 65.

community, that the possessor is a success. In this sense — in revealing what had not hitherto been seen — *The Theory of the Leisure Class* is a major scientific achievement.

Alas, also, much of the process by which this truth is revealed — by which Veblen's insights are vouchsafed — is not science but contrivance. Before writing *The Leisure Class,* Veblen had, it is certain, read widely in anthropology. Thus he had a great many primitive communities and customs at his fingertips. And he refers to these with a casual insouciance which suggests — and was meant to suggest — that he had much more such knowledge in reserve. But the book is devoid of sources; no footnote or reference tells on what Veblen relied for information. On an early page he explains that the book is based on everyday observation and not pedantically on the scholarship of others. This is adequate as far as Fifth Avenue and Newport are concerned. There accurate secondhand knowledge could be assumed. But Veblen had no similar access to everyday knowledge about the Papuans.

In fact, Veblen's anthropology and sociology are weapon and armor which he contrives for his purpose. He needs them to illuminate (and to make ridiculous) the behavior of the most powerful — the all-powerful — class of his time. By doing this in the name of science and with the weapons of science — and with no overt trace of animus or anger — he could act with considerable personal safety. The butterfly does not attack the zoologist for saying it is more decorative than useful. That Marx was an enemy whose venom was to be returned in kind, capitalists did not doubt. But not Veblen. The American rich never quite understood what he was doing to them. The scientific pretense, the irony and the careful explanations that the most pejorative words were being used in a strictly nonpejorative sense put him well beyond their comprehension.

The protection was necessary at the time. And there is a wealth of evidence that Veblen was conscious of its need. During the years when he was working on *The Leisure Class,* liberal professors at the University of Chicago were under frequent attack from the neighboring plutocracy. The latter expected economics and the other social sciences to provide the doctrine that graced its privileges. In the mid-nineties Chauncey Depew told the Chicago students (in an address quoted by Joseph Dorfman) that "This institution, which owes its existence to the beneficence of Rockefeller, is in itself a monument of the proper use of wealth accumulated by a man of genius. So is Cornell, so is Vanderbilt, and so are the older colleges, as they have received the benefactions of generous, appreciative, and patriotic wealth."[16] In 1895, Edward W. Bemis, an associate professor of political economy in the extension, i.e., outpatient, department of the university, attacked the traction monopoly in Chicago which, assisted by wholesale bribery, had fastened itself on the backs of Chicago streetcar patrons. There was a great uproar, and his appointment was not renewed.

The university authorities, like many godly and scholarly men in academic positions, took for granted that their devotion to truth accorded them a special license to lie. So they compounded their crime in dismissing Bemis by denying that their action was to appease the traction monopoly or that it reflected any abridgment of academic freedom. The local press was not misled; it applauded the action as a concession to sound business interest. In a fine sentence on scholarly responsibility, *The Chicago Journal* said: "The duty of a professor who accepts the money of a university for his work is to teach the established truth, not to engage in the 'pursuit of truth.' "[17] A forthright sentiment.[18]

---

16 Chauncey Depew, quoted in Dorfman, p. 122.

17 *The Chicago Journal*, quoted in Dorfman, p. 123.

18 The history of Bemis's discharge is the subject of a recent study by Harold E. Bergquist, Jr., "The Edward W. Bemis Controversy at the University of Chicago," *AAUP Bulletin*, vol. 58, no. 4 (December 1972), pp. 384–393, which appeared after this article originally went to press. While suggesting more complex circumstances than here implied and also a less than innocent role for J. Laurence Laughlin, Mr. Bergquist's conclusions as to the dismissal of Bemis are much as above.

Veblen did not miss this lesson. The last chapter of *The Leisure Class* is on "The Higher Learning as an Expression of the Pecuniary Culture." It anticipates a later, much longer and much more pungent disquisition by Veblen on the influence of the pecuniary civilization on the university (*The Higher Learning in America; A Memorandum on the Conduct of Universities by Business men,* published in 1918.) In this chapter Veblen stresses the conservative and protective role of the universities in relation to the pecuniary culture. "New views, new departures in scientific theory, especially new departures which touch the theory of human relations at any point, have found a place in the scheme of the university tardily and by a reluctant tolerance, rather than by a cordial welcome; and the men who have occupied themselves with such efforts to widen the scope of human knowledge have not commonly been well received by their learned contemporaries."[19] No one will be in doubt as to whom, in the last clause, Veblen had in mind. Elsewhere, contemplating university administration, he notes that "As further evidence of the close relation between the educational system and the cultural standards of the community, it may be remarked that there is some tendency latterly to substitute the captain of industry in place of the priest, as head of seminaries of the higher learning."[20]

Given such an environment and given also his subject, it will be evident that Veblen needed the protection of his art. On the whole, it served him well. In the course of his academic career he was often in trouble with university administrators — it was they, not the great men of industry and commerce, who kept him moving. His more orthodox and pedestrian, though more fashionable, academic colleagues also disliked him. A man like Veblen creates great problems for such people. They cherish the established view and rejoice in the favor of the Establishment. Anyone who does not share their values is a threat to

19 Thorstein Veblen, pp. 245–246.
20 Thorstein Veblen, p. 242.

their position, and, worse still, to their self-esteem, for he makes them seem sycophantic and routine, which, of course, they are. Veblen, throughout his life, was such a threat. But the rich, to whom ultimately he addressed himself, rarely penetrated his defenses.

Veblen also enjoyed a measure of political immunity in a hostile world because he was not a reformer. His heart did not beat for the proletariat or otherwise for the downtrodden and poor. He was a man of animus but not of revolution.

The source of Veblen's animus has regularly been related to his origins. As the son of immigrant parents, he had experienced the harsh life of the frontier. This was at a time when Scandinavians were, by any social standard, second-class citizens. They were saved only because they could not readily be distinguished by their color. What was more natural than that someone from such a background should turn on his oppressors? *The Theory of the Leisure Class* was Veblen's revenge for the abuse to which he and his parents were subject.

This misunderstands Veblen. The Veblens, we have seen, were not of the downtrodden. And as one from a similar background perhaps can know, there is danger of mistaking contempt or derision for resentment. Some years ago, to fill in the idle moments of an often undemanding occupation — that of the modern ambassador — I wrote a small book about the clansmen among whom I was reared on the north shore of Lake Erie in Canada. The Scotch, like the Scandinavians, inhabited the farms. The people of the towns were English. They were the favored race. In Upper Canada in earlier times, Englishmen, in conjunction with the Church of England as a kind of holding company for political and economic interest, dominated the economic, political, religious and social life to their own unquestioned pecuniary advantage.

Our mood, really that of the more prestigious class, was not, I think, different from that of the Veblens. We felt ourselves superior to the storekeepers, implement salesmen, grain dealers and other entrepreneurs of the adjacent towns. We worked harder, spent less, but usually had more. The leaders among the Scotch took education seriously and, as a matter of course, monopolized the political life of the community. Yet the people of the towns were invariably under the impression that social position resided with them. Being English and Anglican, they were identified, however vicariously, with the old ruling class. Their work did not soil the hands. We were taught to think that claims to social prestige based on such vacuous criteria were silly. We regarded the people of the towns not with envy but with amiable contempt. On the whole, we enjoyed letting them know.

When I published the book, I received a quite unexpected flow of letters from people who had grown up in the German and Scandinavian towns and small cities of the Midwest. They told me that it was their community that I had described. "That was how we felt. You could have been writing about our community." The Veblens regarded themselves, not without reason, as the representatives of a superior culture. The posturing of the local Anglo-Saxon elite they also regarded with contempt. *The Theory of the Leisure Class* is an elongation of what Veblen observed and felt as a youth.

The Theory of the Leisure Class, when it appeared, admirably divided the men of reputable academic position from those who were responsive to ideas or capable of thought. One great man said that it was such books by dilettantes that brought sociology into disrepute among "careful and scientific thinkers." Science, as ever in economics and sociology, was being used to disguise orthodoxy. He went on to say that it was illegitimate to classify within the leisure class such unrelated groups as the barbarians and the modern rich. Another predictable scholar avowed that the rich were rich because they earned the money; the gargantuan reward of the captain of industry and the miserly one of the man with a spade were the proper valuation of their contribution to society as measured by their economic efficiency. The names of these critics are now lost to fame.

Other and more imaginative men were delighted. Lester Ward, one of the first American sociologists of major repute, said that "the book abounds in terse expressions, sharp antitheses, and quaint, but happy phrases. Some of these have been interpreted as irony and satire, but ... The language is plain and unmistakable ... the style is the farthest removed possible from either advocacy or vituperation."[21] Ward was admiring but a bit too trusting. William Dean Howells, then at the peak of his reputation, was equally enthusiastic. And he too was taken in by Veblen. "In the passionless calm with which the author pursues his investigation, there is apparently no animus for or against a leisure class. It is his affair simply to find out how and why and what it is."[22] (For these reactions, I am again indebted to Dorfman.) The sales of *The Leisure Class* were modest, although few could have guessed for how long they would continue. Veblen was promoted in 1900 to the rank of assistant professor. His pay remained negligible.

In the years following the publication of *The Theory of the Leisure Class* Veblen turned to an examination of the business enterprise in its social context — an interest that is foreshadowed in *The Leisure Class* in the distinction between "exploit," which is that part of business enterprise that is devoted to making money, and "industry," which is that part that makes things. (In a characteristically matter-of-fact assertion of the shocking, Veblen notes that "employments which are to be classed as exploit are worthy, honorable, noble"; those involving a useful contribution to product being often "unworthy, debasing, ignoble.")[23] In 1904, Veblen developed this point (and much else) in *The Theory of Business Enterprise.* Out of

21 Lester F. Ward, quoted in Dorfman, p. 194.
22 William Dean Howells, quoted in Dorfman, p. 196.
23 Thorstein Veblen, p. 29.

his meager income, he was still required to pay a good part of the publishing cost himself.

In introducing a recent (and handsomely selling) French edition of *The Theory of the Leisure Class*, Raymond Aron argues that Veblen was better in his social than in his economic perception. With this I agree. The basic idea of *The Theory of Business Enterprise* is a plausible one — I can still remember my excitement when I first read the book in the thirties while a student at Berkeley, where the Veblen influence was strong. There is a conflict between the ordered rationality of the machine process as developed by engineers and technicians and the moneymaking context in which it operates. The money-makers through competition and interfirm aggression, and the resolution of the latter by consolidation and monopoly, sabotage the rich possibilities inherent in the machine process. But — though some will still object — the idea has been a blind alley. Organization and management are greater tasks than Veblen implies; so is the problem of accommodating production to social need. And so is that of motivation and incentive. All of this has become evident in the socialist economies, where far more difficulties have been encountered in translating the rationality of the machine process into effective economic performance than Veblen would have supposed. In the thirties, after Veblen's death, the political movement (perhaps more properly the cult) "technocracy" was founded on these ideas by Howard Scott. For a while it flourished. Had the technocrats been given a chance, they would have faced the same problems of management, organization and incentives as have the socialist states. Though much read in the first half of the century, *The Theory of Business Enterprise*, unlike *The Theory of the Leisure Class*, has not similarly survived.

Veblen's writing continued, and so, in 1906, did his academic peregrinations. His classes were small, his reputable academic colleagues adverse and his married life perilous — he was increasingly disinclined to resist the aggressions of admiring women. But in a minor way he was famous and thus a possible academic adornment. Harvard, urged by Frank W. Taussig, who had a recurring instinct for dissent, considered inviting him to join its department of economics but soon had second thoughts. David Starr Jordan, then creating a new university south of San Francisco, was not so cautious. He invited Veblen to Leland Stanford as an associate professor. Veblen survived there for three years. But his domestic arrangements — sometimes Ellen, sometimes others — were now, for the times, a scandal. Once he responded wearily to a complaint with a query: "What is one to do if the woman moves in on you?" Jordan concluded that there were adornments that Stanford could not afford. Veblen was invited to move on. By the students, at least, he was not greatly missed. Though dozens were attracted to his classes by his reputation, only a handful — once only three — ever survived to the end of the term.

After he left Stanford, another established scholar with an instinct for the dissenter came to his rescue. H. J. Davenport, then one of the major figures in the American economic pantheon, took him to the University of Missouri. There he encountered some of the students on whom he had the most lasting effect. One, Isador Lubin, was later to be a close aide of Franklin D. Roosevelt and a protector of Veblen in the latter's many moments of need. Veblen divorced Ellen and in 1914 married Anne Fessenden Bradley, a gentle, admiring woman who did not long survive. In 1918, she suffered severe mental illness, and in 1920 she died. From Missouri Veblen went to Washington during World War I as one of the less likely participants in the wartime administration. From Washington he went on to New York to experiment with life as an editor and then to teach at the New School for Social Research. His writings continued; as were the early ones, they are sardonic, laconic and filled with brilliant insights.[24] As with

24 *The Instinct of Workmanship and the State of the Industrial Arts* (1914); *Imperial Germany and the Industrial Revolution* (1915); *An Inquiry into the Nature of Peace and the Terms of its Perpetuation* (1917); *The Higher Learning*

The Theory of Business Enterprise, many develop points of which there is a hint or more in *The Leisure Class.* The men of established reputation continued to be appalled. Reviewing *The Higher Learning in America* in the *New York Times Review of Books* in 1919, one Brander Matthews said of Veblen, "His vocabulary is limited and he indulges in a fatiguing repetition of a dozen or a score of adjectives. His grammar is woefully defective . . ."[25] The book is, in fact, one of Veblen's most effective tracts. Other reviewers were wiser. Gradually, step by step, it came to be realized that Veblen was a genius — the most penetrating, original and uninhibited source of social thought in his time.

This did not mean that he was much honored or rewarded. Veblen's students and disciples frequently had to come to his support. Employment became harder to find than ever. In the mid-twenties, aging, silent, impecunious and tired, he returned reluctantly to California, and there, in 1929, he died.

The Nation, following his death, spoke of Veblen's "mordant wit, his extraordinary gift of . . . discovering wholly new meanings in old facts,"[26] saying in one sentence what I have said in many. Wesley C. Mitchell wrote an obituary note for *The Economic Journal,* the journal of The Royal Economic Society, then the most prestigious economic publication in the world. Saying sadly that "We shall have no more of these investigations, with their curious erudition, their irony, their dazzling phrases, their bewildering reversals of problems and values,"[27]

---

in America; A Memorandum on the Conduct of Universities by Business men (1918); *The Vested Interests and the Common Man* (1919); *The Place of Science in Modern Civilisation and Other Essays* (1919); *The Engineers and the Price System* (1921); *Absentee Ownership and Business Enterprise in Recent Times; The Case of America* (1923). At the end of his life Veblen resumed an early interest in his Norseland origins and studied the Icelandic sagas. His last publication was *The Laxdaela Saga* (1925).

25 Brander Matthews, The *New York Times Review of Books*, March 16, 1919, p. 125.

26 *The Nation,* vol. 129, no. 3345, p. 157.

27 Wesley C. Mitchell, "Thorstein Veblen: 1857–1929," *The Economic Journal,* vol. 39 (1929), p. 649.

he also observed that *The Economic Journal* had reviewed but one of Veblen's books. In 1925, it had taken notice of the ninth reprinting of *The Theory of the Leisure Class*. This was twenty-six years after its original publication.

12

A Note on the Psychopathology
of the Very Affluent

A SERIOUS LAG exists between the avowed political concerns of our time and the kinds of studies that are being done in universities and other places of solemn thought. For many decades, beginning at least with the thirties, the official concern of the country has been with the poor. In consequence, they have been much studied. Their education, ethnic composition, marital and sexual tendencies, psychiatric afflictions, unemployment and shortage of income have all been subjects of exhaustive academic attention. They still are, and therein lies the lag.

For the official concern of the government, we all know, has now changed. President Nixon made it perfectly clear, to use his words, that those who have asked what they could do for themselves and have found a profitable answer are now the proper object of public concern — along with those whose ancestors asked and answered for them and, additionally, it now appears, quite a few who simply helped themselves. Concern especially for the taxes of the affluent has especially increased. Yet the academic preoccupation remains unchanged. The poor are still being studied. The Ford Foundation is funding practically no work on the rich. It is this situation that the present essay is designed, in some small part, to correct.

I've been studying the problems of the rich under exceptionally favorable circumstances in the village of Gstaad in Switzerland. Partly this is the result of an accident; I started going there to write some twenty-five years ago, and the rich moved in on me. Of necessity, though, my observation has been somewhat at second hand. A scholar who is working on Watts, Bedford-Stuyvesant or the Appalachian plateau can get out with his people. If you are a serious writer, that is impossible with the rich. It is the nature of the wealthy existence that it involves the most elaborate possible waste of time, not that wasting time is unknown in university circles. Some of my colleagues have identified it with academic freedom and raised it to the level of a scholarly rite. However, hearing of my interest, a couple of exceedingly handsome women — one the wife of a motion-picture producer, the other of an Italian automobile magnate — volunteered to help. Both were in a position to waste a great deal of time.

The last great tract on the problems of the rich was *The Theory of the Leisure Class,* published just before the turn of the century.[1] Much, we have discovered, has changed since then. In 1899, wealth, by itself, was a source of distinction. It was necessary only that people knew one had it. Accordingly, Thorstein Veblen wrote of the ways by which the wealthy advertised their wealth — of the methodology of conspicuous consumption, conspicuous waste and conspicuous leisure. Mansions, carriages, clothes and social festivity were the most suitably conspicuous forms of consumption. If carried beyond a certain point of excess, all involved a satisfactory manifestation of waste. Leisure, in a world where nearly everyone had to work to survive, was sufficiently conspicuous in itself. But the point could be driven home by clothing — corsets, hoop skirts, high silk hats — that was palpably inconsistent with any form of toil.

The modern problem of the rich is both simpler and more difficult: wealth is no longer exceptional and therefore no

---

[1] See the preceding essay, "Who Was Thorstein Veblen?"

longer a source of distinction. Yet the rich still yearn for distinction. The problem is exacerbated by their strong tribal
tendencies. They flock and hunt together, and if everyone
around is loaded, money and conspicuous expenditure do even
less for an individual. (In addition to the usual inducements —
the seasons, tradition and the tax authorities — whim appears
to play a role in the migratory tendencies of the rich. My
researchers told me that on a certain day last winter Gstaad
suddenly became unfashionable, and the rich all went to Rio
for the carnival. An aging fellow traveler of the affluent who
was without funds but could not afford to be separated from
the mob took himself, according to legend, to the neighboring
town of Bulle during carnival time and asked a trusted ally to
mail prearranged postcards home from Brazil.) In any case,
last winter a man who lost $100,000 at backgammon in one
evening got almost no notoriety from his outlay and very few
invitations as a result. In an even sadder case, a woman who
combines great wealth with repellent appearance and advanced
nymphomania paid $300,000 for a lover — the technique is to
deposit the money in the local bank and ensure reliability and
durability by limiting the amount that can be withdrawn in
any one month — and got no mileage from it at all, only the
lover. One of my assistants, the wife of the motorcar man, says
she was once propositioned by a twenty-year-old Italian who
simply wanted an automobile. She offered to put him in touch
with the wife of a good used-car dealer.

There is a further problem with the classical forms of conspicuous consumption: they are often positively inconsistent
with the quest for distinction. Thus extra weight and a boozy
appearance, once an index of rank, are now damaging for a
woman and no longer do much even for an Englishman of
noble birth. The average proletarian, after having dined with
the rich, would need to stop on the way home for a hamburger.
Similarly, houses without people to manage them reduce the
owners to work, which is an undistinguished thing. Broadly

speaking, no one in the United States or Europe ever serves anyone else except as a matter of stark necessity. Additionally, houses that are merely expensive are said to lack taste — because they usually do. Something can be done to neutralize the latter charge by hiring a decorator. A local aspirant gets some mileage from having the only house decorated by Valerian Rybar. (It was for a reputed hundred grand.) But with most other decorators there is the problem that one must live with the result. Sometimes, although not often, even the rich are sensitive.

Finally, although it is tough to work, idleness no longer has any affirmative value. On the contrary, it has come to be believed that an idle man is unimportant. If a woman is sufficiently beautiful and has a good figure, she can survive idleness, for it is taken for granted that she has ways of occupying at least part of her time. But this role also now invites criticism.

So a person must be both rich *and* distinguished, and distinction is something that money will no longer buy. To be rich and commonplace is to live on the edge of despair. There are the tribal dinners, cocktail parties, gay informal luncheons, receptions for visiting movie producers, stars or directors, and the undistinguished remain at home. They essay festivities on their own, but except for a few characters of deficient wit who are said to be getting by *on* their wits, no one shows up.

Meanwhile an effort to cultivate an aspect of importance encounters grave natural handicaps. The local sample of the rich includes a number of individuals whose families, former husbands or business firms consider it highly advantageous that they live at the greatest possible distance. That is usually because they lack intelligence, charm, emotional stability or any other known attainment, including the ability to read without undue movement of the lips. And, quite generally, the merely rich lack the ability to command the favorable attention they crave, and the ability to do so disintegrates further with age. One of my researchers says firmly that the average rich man has only one chance to excel these days: "He's got to be a real clown." To fend off age, a fair number even resort to a local clinic, where they are injected with cells said to be superior to their own that are supposed to keep them young and virile. However, my other researcher is bearish, or certainly not bullish, about this: "The most it's ever done for any man I've known is to give him a sore ass." In addition to the cells, there are fraudulent pills to prevent aging supplied to a major concentration of customers who want them by the two drugstores in Gstaad.

From the foregoing, it will seem that the affluent are now not nearly as happy as the conspicuous and uncomplicated rich of earlier times. The past year, however, may have altered things a bit. Some of the rich, oddly enough, have had their neuroses subsumed by old-fashioned worry about money. In Gstaad a distinction is made between the rich and "the only two-house rich." The latter, who may also be refugees from alimony or the IRS, spend pretty much all of the money they get. For many, income comes in dollars. The several devaluations and slumps have had a marked effect on people's personal economy. Quite a few of the afflicted have stopped me in the village to ask my views on the monetary situation; and twice couples have crept into our apartment to inquire. One man, with a look of woe I haven't seen since our troops overran Dachau, said he might even have to go to work.

I endeavored to help by telling my patients, if they were Americans, that they should count on the dollar's going to zero or perhaps a trifle below. As I developed this thesis, I could see a different look — that of anxiety — spreading over the leisure-ravaged faces. I knew I had rescued fellow humans from the deeper anxieties of the rich and returned them to the simple, old-fashioned, manageable worries about money that everyone else has.

II
Personal History

13

My Forty Years with the FBI

The graduate students with whom I associated in the thirties were uniformly radical and the most distinguished were Communists. I listened to them eagerly and would liked to have joined both the conversation and the Party but here my agricultural background was a real handicap. It meant that, as a matter of formal Marxian doctrine, I was politically immature. Among the merits of capitalism to Marx was the fact that it rescued men from the idiocy of rural life. I had only very recently been retrieved. I sensed this bar and I knew also that my pride would be deeply hurt by rejection. So I kept outside. There was possibly one other factor. Although I recognized that the system could not and should not survive, I was enjoying it so much that, secretly, I was a little sorry.

I WROTE THE ABOVE nine or ten years ago for a volume celebrating the centenary of the University of California, to which I had proceeded for graduate study in 1931 after taking a degree in animal husbandry at the Ontario Agricultural College. I am able to reproduce the item without going back to the manuscript, for it appears in my FBI file under date of September 17, 1971, just forty years after my original temptation. Government employment was not involved; on the undesirability of that at the time Mr. Nixon and I were in an unnatural agreement. Rather, I was about to be elected President of the American Economic Association, an honor often associated with longevity, and a member of the association had written to J. Edgar Hoover, enclosing the foregoing confession and saying that while he did not expect any action, he did want Mr. Hoover to know that "the trend is of concern to many in the profession." The Director, who had much such help, replied six days later with a three-line letter of thanks, beginning "Dear Dr." and carrying a total of, possibly, ten different initials according clearance.

This mild item is included in my FBI file, the most interesting set of government documents that has ever come my way. This file, good for several days' reading, cultivates a sense of deep paranoia, less one's own than that of the large number of one's fellow citizens who live in fear of Communists and Communism and the even larger number who once desperately feared they would be thought soft on Communism and in consequence be heaved out of their jobs. It tells, also, how difficult it was to decide what qualified an individual as a Communist or a dangerous radical or as being otherwise inimical to the system. Dubious personal traits, even a badly exaggerated ego, might serve. The file is unparalleled in my experience as a mine of misinformation. Also it proves conclusively that on the matter of being a security risk — perilous one day, safe the next — the age of miracles is not over.

While the impression of other people's paranoia is great, my own was diminished by the fact that while the documents are full of deeply damaging intentions, virtually nothing unpleasant ever happened as a consequence. (But one can see how the only slightly more vulnerable must have suffered. It is good to be on even the raffish fringe of the Establishment.)

The file also proves, and here beyond the most pallid shadow of a doubt, that the government of the United States has, in these matters, a colossal capacity for wasting money. Some tens of thousands of old-fashioned real dollars were spent in 1950 investigating my fitness to continue in a job in which I had rendered no service and of which I was unaware until the investigation culminated one day in a demand that I tell all details of my association with Mr. Corliss Lamont. Mr. Lamont, a neighbor, friend, radical, civil libertarian and son of a Morgan partner, was considered an especially dangerous companion for anyone employed without his knowledge in a nonexistent job. On this, as on other matters, there is much that is very funny. There is also much that evokes one's sympathy, even admiration, for the rank-and-file member and down-the-line agent of the FBI struggling with these pathetic tasks. But let me begin at the beginning.

My first jobs with the United States government involved only the most benign of relationships with the FBI, and at the outset none at all. In June of 1934, on the way from Berkeley to Harvard, I stopped over in Washington and was promptly put on the payroll for the summer by a former professor as an associate economist in the Agricultural Adjustment Administration. Economists were in short supply. The AAA had, during this period or just before, enrolled a number of radicals of later fame, including Alger Hiss, Lee Pressman, John Abt, Nathan Witt, Nathaniel Weyl as well as Jerome Frank, George Ball and Adlai Stevenson. As with the Berkeley radicals, I never at the time achieved the distinction that allowed me to know any of them.

In those days one went on the payroll without FBI clearance, the FBI being generally regarded as a law enforcement agency, and I don't remember that I was even asked if I was a citizen, which I was not. Clearance was, however, required from James A. Farley, custodian and dispenser of Democratic patronage. His representative had an office on the top floor of the South Building of the Department of Agriculture, and he called me up and made me affirm that I was a Democrat. This I did with good conscience. In southern Ontario everyone adored Roosevelt, and certainly no one at Berkeley had been for Hoover.

I worked further on various public tasks in the next few years without being aware of the Bureau or vice versa. This changed in the summer of 1940, when I was employed by the

Advisory Commission to the Council of National Defense. Elementary investigations were ordered to exclude spies from such posts. The resulting reports can be read with nearly undiluted admiration of both one's self and of the investigators. At Berkeley, Cambridge, Princeton and the American Farm Bureau Federation in Chicago, all places of my previous study or employment, the agents were told, and faithfully reported, that I was "brilliant," an "excellent writer," possessed of "a keen sense of sportsmanship" (something of which I was not then or thereafter aware), of a "good personality," "not obnoxious," a "good conversationalist" and with no adverse credit record, in fact none at all. It was said, no one being perfect, that the subject "was impressed with his own knowledge and importance" and "was too deep a thinker for undergraduates." Also, "a poor public speaker."

From those very earliest days one detects a tendency, highly developed in all later investigations, for one's friends to sense with precision what statement would be the most damaging to one's public career and then to volunteer, with great emphasis and some talent for invention, the precise opposite. Eventually there were to be numerous (by FBI standards) derogatory items in the file, to which I will come, and while the good things disappear, the bad live on. Gresham's Law operated relentlessly here. But more than half of the file by volume consists of extravagant attestations to whatever quality would most allay suspicion. Thus, during the 1940 investigation, the longtime Chairman of the Harvard Department of Economics — a deeply conservative scholar of modest attainments named Harold Hitchings Burbank (he is easily identified from the context) — was forced to concede, as a matter of simple intellectual honesty, that "the subject leaned as far to the left as President ROOSEVELT," who was then in office. But then he moved quickly to retrieve. I was extremely loyal. I also had a fine military aspect — "commanding appearance due to his height of 5'6" [I am 6' 8½"] and his dignified bearing." Another

Harvard professor went further to urge that I "was a conservative thinker and talker," and a Berkeley academician went all out and described me as "reactionary" and therefore "entirely desirable from every angle." One agent did pick up word that I was currently in Cuba having a nervous breakdown. My nerves at the time were fine; Cuba, although then a thoroughly respectable place for a holiday, I had never seen. The misinformation begins at the beginning.

Needless to say, I was cleared, and in the next few years I had even more reason to love the FBI and J. Edgar Hoover, for there was the small matter of a murder rap. One day in late 1941 or early 1942, I arrived at the OPA offices (I had been put in charge of all prices in the United States without any evident investigation) to find two staff members waiting for me, their faces gray with anxiety. A few weeks before, the Navy, a major consumer of sponges for some arcane shipboard reason, had complained about the prices it was having to pay. The two men had gone to Tarpon Springs, Florida, an acknowledged center of the industry, to hold hearings prior to setting a ceiling. The first hearing came to a violent end, the violence having been provoked by the local leader of the sponge fishermen, a man of Greek antecedents and forthright reputation, named — on this, invention replaces history — Nikolas Bolenkus. Further hearings were called and attended by no one because Nick's men were patrolling in a menacing way outside. Eventually Nick called in at the hotel to suggest, helpfully, that our chaps might just as well go back to Washington; they were accomplishing nothing in Florida. In the presence of numerous witnesses, one of my men told Nick that he was about to meet with a major misfortune at the hands of another individual who was bigger and just as tough as he. This man, he said — combining emphasis with imagination — was his boss, J. K. Galbraith. It was legal disaster that he had in mind, but that was not made clear. The threat delivered, the two price-fixers left for Tampa and the plane to Washington. The evening before our meeting in the office, someone had called them from Florida. Earlier that day Nick had stepped on the starter of his automobile to which someone had wired a very large charge of dynamite. Both the car and Nick were totaled. My men then remembered their threat.

I recall thinking at the time that my alibi was better than my worried friends imagined. But I put in a call to one of Hoover's acolytes at the FBI and told him the situation. Toward noon Edgar, as, believe it or not, we called him in those days, phoned back. A cursory inquiry had revealed that I was twenty-ninth in plausibility (that is my recollection of my rank) as suspected murderer among those who had been heard threatening to knock off the late Bolenkus. My men were relieved, and I had achieved a story which I've since told a hundred times. Not every Harvard professor has been involved, however remotely, in a gangland killing.

Over the next nine years my relations with the FBI remained pleasant and also fragmentary. In the closing days of the war and for some time thereafter I was myself involved in investigatory activities. In 1944, President Roosevelt, having, as I've often said, mastered the first principle of modern warfare, which is that the claims of air generals as to what they are accomplishing have no natural relationship to truth, asked that a special study group — the U.S. Strategic Bombing Survey — be constituted to establish the facts. Others, including George Ball, were urging the same need. I became a director the following year, but this involved no new investigation of my background. However, I did become knowledgeable on economic conditions in Germany and Japan, and this led to my being given charge of economic affairs in these countries (as well as Korea and Austria) in the State Department in early 1946.

My years as a price-fixer had been richly controversial. I was thought to enjoy severity for its own sake, which may have been true, and "radical theorist" was the term of opprobrium on which all of my critics eventually converged. My resignation from the OPA in 1943 had provoked more conservative applause than I have achieved since. In consequence, the Civil Service Commission and the special security investigators of the State Department were moved to investigate at the time of my new appointment. The State Department was not, in those days, a nest of radical theorists.

The job was not one I enjoyed. General Lucius Clay in Germany was not impressed with my guidance on economic affairs, and General Douglas MacArthur in Japan may not have been aware of it. The investigation as to my suitability for the task was not completed until after I had concluded, in the autumn of 1946, that I wasn't being very useful and had left.

These investigators turned up the usual and numerous encomiums on my loyalty and conversational tendencies, and on my loyalty these were as eloquent as before. However, now there was some bad-mouthing from people whose prices I had fixed or those who disapproved of fixing on principle. Congressman John Taber, an articulate fossil from upstate New York with whom I had often clashed, told a Civil Service investigator that Galbraith "was fired by the President and the Board of Trustees from his job as Professor at Princeton University because he was a Communist. He is a member of many Communist front organizations . . . a Totalitarian . . . would be a whole lot more effective using a pick and shovel . . . [in the State Department] would absolutely be a menace." And someone I judge to have been an aged Princeton professor of economics advised the investigator that I was pretty doctrinaire in my views, "in favor of anything Russia was in favor of." The reference to Russia did not survive, but his description of me as doctrinaire was a near-catastrophe, for it was heard as "Doctorware" by the investigator and was held to imply that I was a follower of an otherwise unidentified subversive called Doctorware, later promoted, academically, to "Dr. Ware." For the next twenty years, whenever my file was examined, the superb testimony on my personality, garrulity and loyalty was never reproduced. Only the references to radical theory, to Dr. Ware and to the action of the Princeton President and Trustees. In time, the FBI, having come into possession of the Civil Service files, sent an agent back to interview Taber, and he then denied all knowledge of my discharge from Princeton. (As the university frequently advised the Bureau, I had been an assistant professor, and my three-year term had expired while I was on leave with the OPA.) He also denied all other firsthand knowledge of my life and loyalty. He passed the FBI on to his own source, an "investigator" for the Republican National Committee, and *he* denied all knowledge of any kind. Still, the impression remained permanently in the file that there was something very funny about my departure from Princeton. Perhaps it was thought that the Communists had somehow got to those who said it was routine, although here I am just guessing.

Another durable piece of information from these years came from a newspaper clipping. During the war it was charged that my controls on newsprint prices were drying up the supply. This a congressman had publicized as proof that I was a member "of a group that participated in 'an effort to curtail drastically the amount of newsprint available to [the] free press.'" This charge also survived and, indeed, was never disproved.

In consequence of the foregoing information, or such as was by then available, and the controversy over my price-fixing, the Security Screening Committee of the Security Office of the Department of State on January 25, 1946, formally disapproved my appointment as a security risk. It also concluded that my being on the payroll would "draw sharp criticism of the Department . . . and . . . jeopardize as a result . . . certain programs and appropriations." It concluded that "It cannot be conceived that this applicant possesses qualifications which will in anyway [sic] offset or compensate for the resulting damages to the Department's prestige." After this finding, I was promptly and routinely appointed and did not know of the interdict until I got the file. The investigatory routine had already become silly. During the war I had worked closely with James F. Byrnes, who was now Secretary of State, and W. L. Clayton, his Assistant Secretary for economic matters. They knew me well, and it was natural that they would ignore investigators who did not.

It was in 1950 that my relations with the Bureau became really intimate and detailed. Of this I was also unaware. In 1948, I left *Fortune* magazine where I had been an editor — *Fortune* had only a moderately more subversive reputation then than now — and returned to Harvard. During the summer of 1950, I was in Europe on vacation with my wife and son, and in Switzerland one day I received an urgent telephone call from the Economic Cooperation Administration in Washington — the back-up organization of the Marshall Plan — asking if I would go to Frankfurt and Bonn and work out arrangements involving a joint German-American commission to examine the refugee problem, a matter on which I was deemed to have some special competence. It was several days before travel clearance arrived, and later, when I was back in Washington, I asked the man who had called me why, after all the urgency, all the delay. He told me that it was a time-consuming task to read my security file and that the man who had started on it had been transferred to another job before he finished. During those summer days the file was, in fact, growing at a spectacular rate.

Earlier that year — on the 24th of February to be exact — I had attended for a few hours a meeting at the Department of Commerce in Washington to consider the effect of the agricultural subsidy programs on the economy. I had filed various forms to claim travel and compensation, matters I have never been inclined to neglect. Unknown to me, one of these had put me on the rolls of the Department of Commerce as a consultant when, as and if employed — which, since I was not again employed, was not at all. This, in turn, made me subject to the deepening concerns over the LOYALTY OF GOVERNMENT EMPLOYEES, as the investigative forms were headed — a concern then gathering force in response to the trials of Alger Hiss and the fear of Senator Joseph R. McCarthy. A preliminary check by the FBI turned up the alarming references to radical theory, Dr. Ware (still called Doctorware), the righteous if imagined action of the Princeton President and Trustees, the conspiracy against the free press and a couple of items of real if less than subversive substance. In 1941, at a congressional hearing, I had come to the support of a onetime Berkeley professor of mine, Robert A. Brady, who was under heavy fire for having had a book distributed in England by the Left Book Club. I had assured the committee that the Left Book Club, which specialized exclusively in works from well left of center, was the English equivalent, more or less, of the Book-of-the-Month Club. There was a large element of fantasy here; there is always a temptation in such hearings to say whatever will tranquilize an aroused committee and then get out of the room. You never should. Further, in 1944, I had been active in the National Citizens Political Action Committee, a body organized by Sidney Hillman to work for the reelection of President Roosevelt. It unquestionably enrolled some very active Communists, an association which I'm glad to say I did not then, and have not since, believed destructive. And we had all been for the reelection of Roosevelt. Although the NCPAC was not one of the proscribed organizations of the Attorney General or even of the House Un-American Activities Committee, it had fallen under the ban of the relatively much less discriminating California Committee on Un-American Activities. Their list of subversives and subversive organizations had a kind of cadet standing and was regularly reviewed by Washington. All this was enough to cause the Department of Commerce to ask the

Civil Service Commission to request the FBI on March 28, 1950, to convert the superficial check into what the FBI calls an FFI — a Full Field Investigation. Frightened officials in Commerce and not the FBI, it should be noted, were responsible.

It was a very full investigation indeed, and it was this that must have run into the real money. Men were deployed, according to a later memorandum, in Washington, New York, Boston, Chicago, Newark (meaning Princeton), Newark again, Detroit, San Francisco (meaning Berkeley), Chicago, Richmond (meaning suburban Washington), Richmond again, Birmingham, Albany, Boston again and St. Louis. A request went to the State Department for research, via the consular offices, into my Canadian background and my activities while a student thirteen years before in England. "Should substantive information be developed [in these countries] reflecting disloyalty on the part of the Appointee, it would be appreciated if signed statements are obtained from persons furnishing such disloyal information." No disloyal information, so attested, seems to have been forthcoming. My book reviews were also read by a scholarly agent at the New York FBI office and were found not to "reflect any information that may be pertinent to this investigation." The single exception was a review of Merle Miller's *The Sure Thing*, a novel having to do with the current witch-hunting, and this was sent as an enclosure with the New York agent's report.

The same agent noted that "The Biographical Morgue file of the 'New York Times' newspaper on the appointee was reviewed by the writer. The file did not reflect any information reflecting on the loyalty of the appointee to the United States."

Such precise language is a passion of the FBI. Summers during the war years we took a house in suburban Virginia on the grounds of an Episcopalian seminary there. The agent reporting on this residence advised the Bureau that "Other efforts to locate persons residing in the vicinity of the appointee's former address on Seminary Hill, Alexandria, Virginia, who were residing there during the appointee's residence, were met with negative results."

In each of the several cities, Communists of known reliability who were operating in the Party on behalf of the FBI were visited.

> Confidential Informants New York City T-2, T-3, T-4, T-5 and T-6, who are reliable and are familiar with general Communist activities in the New York City area, advised that they were not acquainted with the appointee.

Also a helping hand right at Harvard:

> Boston Confidential Informant T-1, of known reliability, closely associated with activities at the school of Public Administration, Harvard University, advised having been acquainted with Appointee... T-1 expressed the firm belief that GALBRAITH was without question a loyal American citizen.

To such of these subversives as survive, my warm thanks.

As before, the FBI agents were overwhelmed with testimony on my loyalty. The speeches were even more extravagant than in 1946 but now with a difference. Before, the agents had heard how affirmatively loyal I was; now they heard how negatively anti-Communist I was. Before, the testimonials had led me to wonder how I escaped high office in the American Legion; now they conveyed the clear impression that I was in hard training for service on or with the HUAC.

There were, however, some nasty notes. A Detroit advertising man named Lou Maxon with whom I had clashed in OPA, after first stressing his personal dislike in a very decent way, "described Appointee as a 'Fly-by-night' economist who seemed determined to inject a Socialist trend in policies and directives of the Office of Price Administration." A Washington agent reported that Mr. J. B. Mathews, research director of the House Un-American Activities Committee and a formidable opponent of Communism and syntax, had testified that "J. KENNETH GALBRAITH has had a connection with one of the Com-

munist books, magazines and other literature but it is not indicated as to exactly [with] which publication GALBRAITH
was affiliated." That book, magazine or other literary affiliation
could only have been with *Fortune*. Other informants who had
suffered under my management of wartime price control or
didn't like it on general principles also got in their licks:
"Screwball in economics" was one of the milder phrases from
a witness, who added that he wouldn't say I was a Communist
"but more of a fellow traveller." Here, however, and with no
nonsense, I must again put in a good word for the Bureau.
On August 21, 1950, as the FFI was getting under way, a
memorandum was sent from Washington to all relevant offices,
advising that my administration of price control during the
war had been viewed with distaste by "many people in this
country, principally members of Congress and business and
industrial leaders" and that I had become a "very controversial
figure." It then went on to say:

> This is being brought to the attention of all offices conducting
> this investigation because it is entirely possible that some witnesses
> may be inclined to give adverse information concerning GAL-
> BRAITH because they were not in agreement with his economic
> theories and policies and such testimony may be given intention-
> ally or otherwise in such a manner to bear adversely upon his
> loyalty. It is therefore suggested that all offices be alert in securing
> testimony in this investigation because of GALBRAITH's back-
> ground.

That was handsome, and I am led to remind the reader once
more that this particular act of investigatory nonsense was in-
stigated not by the FBI but ultimately by the President of the
United States and immediately by the Civil Service Commission
and the Personnel Operations Division of the Department of
Commerce on behalf of the Commodities Division, Office of the
Director, Office of Industry and Commerce, U.S. Department of
Commerce. These are to be blamed. To any of these who also
survive, a vulgar gesture.

The investigation ground on. On October 19, there was a

chilling note. James M. McInerney, an Assistant Attorney General now lost to fame, sent to the Civil Service Commission for the whole file "in connection with this Department's consideration of the above entitled case from the standpoint of possible criminal prosecution under Title 18, Section 1001, U.S.C." This provision of the Code punishes people who lie to federal officials. Leavenworth did not beckon. The request seems only to have been a form letter used to keep the Justice Department in touch with investigations and prepared to act in case those investigated did a snow job on the agents. I was safe; having been asked nothing, I had not lied.

Eventually in late December I did become witting, as the CIA puts it. I received a letter in Cambridge from the Loyalty Review Board of the Department of Commerce, asking, in slightly peremptory fashion, that, as an employee, I disclose my relationship with three men, one of whom I did not know; one of whom, E. Johnston Coil, was my closest friend; one of whom, inevitably, was Corliss Lamont. They wanted also to know about any membership in "subversive" organizations. I answered — friends were friends, of dangerous organizations none. (That parsimony was a restraining factor I did not admit.) Then I asked how come? I wasn't employed. My statement that I didn't hold the job was promptly accepted as a resignation from the job I didn't hold. The investigation, though incredibly still incomplete, was brought to an end. Not quite, in fact. In ensuing years the files kept turning up the fact that I had resigned before my loyalty was established. That was slightly bad.

During the Eisenhower years the risk of even unwitting nonemployment by the government was minimal, but this did not keep me out of the files. The Republic could be threatened in other ways, and my best effort involved a plot to collapse the stock market. This was accomplished one day in March 1955, when I testified before the Senate Committee on Banking and Currency on current conditions in Wall Street and a mini-boom that was then in blossom. During my testimony the market slumped — a total of some $7 billions in values was lost or, as some would have preferred, confiscated. There was a memorable headline, EGGHEAD SCRAMBLES MARKET, and Walter Winchell went on the air to warn Senator Fulbright, then the Chairman of the Banking Committee, that I had been a member of the National Citizens Political Action Committee. "This outfit, Senator, is listed by the House Un-American Activities Committee as a 'Red' front." (Winchell had the wrong committee, but he was not given to precision on such details.) Fulbright was deeply unimpressed, but the message from Winchell did get through to Homer Capehart, Republican of Indiana, then ranking minority member of the committee. He hadn't been around the day I testified; now, on television, he demanded that I come back and explain the plot. He cited a pamphlet on postwar reconstruction issued by the National Planning Association (an upright organization of businessmen, farmers, trade unionists and professors that still functions) in which I was alleged to have said something agreeable about Communism. It was not unrestrained praise; it couldn't have been, for my thoughts had been endorsed by Allen Dulles, by then head of the CIA, and by Milton Eisenhower, a friend from my agricultural days and the brother of the President. I had heard that Capehart was going to unleash and had warned him that a reading of the document would show that it did not serve his purpose. However, though a generally pleasant man, he was not unduly literate and was further handicapped by being deeply obtuse.

When Capehart's attack came, I was prepared. I shouted back with some vigor, and a day or two later, while attending a meeting at Purdue University, I questioned whether anyone so uninformed as to my views should be allowed to represent the people of the state of Indiana. I noted, also, that the contents of the pamphlet had first been given as a lecture at Notre

Dame. This made the Senator guilty by association of an attack on the leading Catholic university in the country and a monument to culture and football in his own state.

I sensed at the time that Capehart was struggling. I was right. The files show a desperate appeal for help. The CIA refused to assist, but Sinclair Weeks of Massachusetts, the Secretary of Commerce, rallied to the Senator's side and asked the FBI for help. A grudge was involved here. Weeks, known to his friends as Sinnie, had taken strong exception to my earlier economic views, and, in reply, I had publicly noted his resistance even to the Renaissance. J. Edgar Hoover, who may, perhaps, have anticipated the Senator's need and Weeks's request — the timing here is difficult to establish — sent a *Washington Post* clipping to his men with the question, "What do our files show on Galbraith?" There followed a frantic scramble for adverse information. "At approximately 4:30 P.M. today . . . I talked to XXXXXX [the X's mean the name is deleted in the file] . . . At 5:50 P.M. I called XXXXXX. Special Memo Section complete a review of all references to Galbraith during the evening of 3-9-55 . . . at 8:50 A.M. . . . I contacted XXXXXX." The contact was with Capehart or Capehart's contact, and the information he or it passed must have been a sore disappointment to Homer. The pamphlet that had seemed to him subversive could not be found in the Bureau's files. It was noted that I had twice been investigated by the FBI. Of the first effort it was said "Investigation favorable except conceited, egotistical and snobbish." This was not favorable, but it was also not the kind of thing that would surprise a United States senator. The second investigation — the FFI — had yielded principally the fact that I had resigned from that nonemployment at Commerce before my loyalty was fully established.

Weeks's request and Hoover's help were highly improper, but once again the FBI was more misused than misbehaving — misused this time not by frightened bureaucrats such as those in Commerce but by a Cabinet member and its own Director. They were the ones to blame. Blame attaches also to President Truman, who tried to protect himself from right-wing criticism with these insane investigations — no other democracy needed them — and later to Presidents Kennedy and Johnson, who should have retired J. Edgar long before God came to the assistance of the Republic.

The files in these years show a more important Hoover aberration that has not, I believe, been celebrated previously. As in the campaign four years before, I served on the speech-writing staff of Adlai Stevenson in 1956. In October of that year, this came to the attention of one of Hoover's ever-vigilant volunteer informants; he wrote urging and very nearly demanding that ghostwriters for candidates be subject to a proper measure of surveillance. ". . . for some time it has been the custom to assign Secret Service men to protect the person of both the principal Presidential candidates in presidential election years. I believe that this procedure should be broadened so as to protect not only the bodies of the candidates, but their minds as well . . . if a president [sic] has not the wish, nor the ability, to put his thoughts into his own words, the 'ghost writer' becomes someone of enormous power . . . It is of the utmost importance to the nation that the 'ghosts' be 'above reproach' like Caesar's wife . . ." I was one of the Stevenson ghosts who did not, in this patriot's view, come even close in purity to the late Mrs. Caesar and was, in fact, one who sent "chills down the spine of any American with a knowledge of the left wing conspiracy to take over our Republic."

There was, of course, some comparative logic in this concern. Were one out to get the free enterprise system and had a choice, one would write speeches for a presidential candidate rather than be an unwitting and nonemployed employee of the Commodities Division, Office of the Director, etc., etc., of the U.S. Department of Commerce. However, on these larger matters, Hoover moved more cautiously. He wrote to the Attorney

General: "I am transmitting herewith a copy of a communication I have received from XXXXXX who suggests that steps be taken to make available to XXXX [Adlai Stevenson] any information pertaining to the background of his alleged 'ghostwriters' . . ." He went on to say that he had acknowledged the letter, "pointing out that this was not a matter within the purview of our responsibility, and I have advised him that I am calling his letter to your attention."

By 1960, however, Hoover had enlarged perceptibly his purview of his responsibility. In that year I was working (though not particularly as an alleged ghostwriter) for John F. Kennedy. On July 5, taking note of this association, Hoover called for a full survey of the files. With the commendable promptness that the Director's wishes produced, five pages of inspired misinformation were on his desk the very next day. The dishonorable discharge from Princeton was there, although by now it was subject to the aforementioned doubts. The Commerce Department was now reported as saying that, in my nonemployment there, I had been viewed as one of 51 "poor security risks," and my departure had been upgraded to a precautionary act — a "voluntary resignation" had been obtained.

Hoover was also told that I was associated, as a Kennedy helper, with Arthur Schlesinger, Jr. There was mention again of Dr. Ware, and it was noted that in 1959, at the suggestion of Adlai Stevenson "with whom he was associated during 1952 and 1956 Presidential campaigns," I had "contacted" the Soviet Embassy. I *had* been associated with Schlesinger even more closely than with Dr. Ware; the approach to the Embassy (unless for a visa) was news.

One learned also what fine distinctions the Director could handle when it came to political views. He was told that mine were " 'left of center,' but not 'left wing,' 'pink' or 'leftist.' " And from the ever-present and decent civil servant came the redeeming note: "Many who disagreed with his economic theories were insistent that while Galbraith's views were 'left of center,' Communism or Socialism could and should not be imputed to Galbraith." The FBI and Hoover had no business so concerning themselves with politics.

While my file does not show that my subversion ever, in the end, kept me off a public payroll, there was some modest pecuniary damage in these years. On one or two occasions people in the CIA asked that I be invited to lecture to the "intelligence community." This was denied by higher authority on the grounds that I was a grave security risk, made worse by the danger that my instruction might provoke criticism from those on the Hill who disliked my views. I was also disapproved for a covert operation, this being a glorious convocation of liberal intellectuals in Milan in 1955 — Hugh Gaitskell, Roy Jenkins, Anthony Crosland, Arthur Schlesinger, George Kennan, many others — which the CIA was secretly funding. (We were told it was funded by a foundation.) But something went wrong here, for, in fact, I attended. After I published *The Affluent Society* in 1958, lower echelons of the United States Information Agency asked regularly for my services as a lecturer and for the book for their libraries. These requests too were firmly refused by more responsible authority. The risk to security and of political criticism was intolerable, although again there was a slip, for I remember giving a lecture under such auspices in Rome.

In this prosaic and excessively intellectual age, there are men and women who do not believe in miracles. Let all be clear: miracles of biblical magnitude still occur. At one point in time, as it is now said, you cannot give a lecture to the CIA or for the United States Information Agency; you are too grave a security risk. At the next point in time, twelve months later, you can be responsible in a vast country for what those agencies do. No ceremony of purification or trial of epuration is involved. Only the continuing marvel of democracy. All this the history now proves.

The earlier investigations were prelude to the biggest investigation of all — that of a putative ambassador. I knew, of course, that this was in progress. It occurred after I had moved into the White House in January of 1961, a fairly strategic location, one would think, where security matters are concerned but one that required no investigation of any kind. Indeed, so far as the files show, the FBI seems never to have discovered that I was there. One day, while my appointment to India was pending, I ran into Adlai Stevenson, who told me he had just been quizzed at length about my loyalty. That impressed me, for earlier on the very same day I had been asked about his. I had told the agent, who was very pleasant, that were Stevenson a subversive, it would rank as one of the more dangerously delayed discoveries of all time.

This investigation revealed another striking fact about the loyalty of government officials. If you are a member of the administration and about to become an ambassador, things go better. Adverse information disappears or even becomes favorable. Thus the Princeton discharge disappeared. So, at long last, did Dr. Ware. My relations with Commerce were reexamined, my letter explaining that I had never been employed was unearthed, and my candor in admitting to my questionable friendships became, I would judge, a plus. All mention of the voluntary/involuntary separation from Commerce before loyalty adjudication vanished. Instead it was noted with emphasis that President Truman had bestowed on me the Medal of Freedom for "exceptionally meritorious achievement" during the war, although, I discovered for the first time, "without palm." I had never missed the palm. My wife's family was now discovered and cited as "of fine character, conduct and reputation and loyal Americans." One or two critics complained that I was "inexperienced in business," and there was, of course, the customary misinformation. My birthplace was given as Ottawa (it appears elsewhere as Toronto and a town on the Detroit River called Sandwich where I have never been and

which has since disappeared). I was described as deeply anti-
Communist, which I am not; it was alleged that I sometimes
said "no" in a tactless fashion when, in fact, I have difficulty
saying it at all. But the errors, like the slurs, were lost in the
massive wave of applause. After noting, among other things,
that I had been described as "a great national figure of unquestioned ability," the FBI became sated and concluded its report
with the truly breathtaking statistic that "ninety-eight other
persons were interviewed and commented favorably concerning the character, reputation and loyalty of Mr. Galbraith.
They also highly recommended him for a position of trust and
responsibility with the Government." The investigatory language, however, was as careful and stately as ever. A Boston
report, dated February 15, 1961, advised that ". . . personnel,
Reference Libraries, Boston Herald-Traveler and Boston Daily
Globe Corporations, both firms which publish newspapers on
a daily basis at Boston, Massachusetts, made available information in the name of appointee, which has been utilized during
this investigation."

The final report was made on March 6, 1961. A week or two
earlier I had heard in the Washington rumor underground that
my appointment was in deep trouble on Capitol Hill — on security grounds. Bourke B. Hickenlooper, a Republican senator
from Iowa and a devout, articulate and loquacious but not especially malicious defender of the system, was standing firm
against me. He had learned that the State Department had
once denied me a passport. An ambassador without a passport
could never get on intimate terms with the leaders of the country to which he was accredited and could well be a nuisance
around Washington. Hickenlooper wanted yet another FFI.
President Kennedy told me he thought the whole business de-
grading. Then suddenly the clouds cleared. Hick, as he was
called by numerous colleagues and constituents, had been ap-
peased. I was puzzled about the original charge, for I had never
been denied a passport. This, at long last, my file explained;

it was only a slight problem in nomenclature, which anyone should have understood.

On February 23, 1961, an FBI agent, checking into things at the State Department, reported back that the files there "disclosed that JOHN KENNETH GALBRAITH was refused an American Passport on 2/20/53 because he was a member of subversive organizations, based on a communication from FBI dated 12/19/51." The same report showed that shortly thereafter I was issued a passport. The first but not the second fact had been sent by some helpful soul to Hickenlooper. A few days later an agent went back to check again. "It was determined by SA [special agent] that this refusal notice does not signify or imply that passport was refused, it is a misnomer, and merely serves as an administrative lookout notice for proper routing of mail within the Passport Office." Anyone should have known this. I went off to India.

There I found that *The Affluent Society* and my other books were still on the Index. They could be risked in libraries only with the special permission of Washington. Few acts of my life ever gave me such a feeling of righteously exercised power as the step I now took to declare my own writing safe for general use.

My association with the FBI had now passed its peak, but it was a long while (in the cost of Xeroxing the file, another ten or fifteen dollars) in decline. While back from India in 1961, I appeared on "Meet the Press" and was asked by one exceptionally handicapped reporter if I thought (as did Nehru) that India should deal with the Russians and Americans on the same moral plane. I said no and observed that an affirmative answer would endanger my security clearance. One of Hoover's volunteer helpers wrote the President in distress — "any loyal American would answer with an unqualified no," and he sent the transcript to J. Edgar for action. Hoover passed. However, in these months another unidentified but more persistent patriot in Birmingham, Alabama, went to the local office of the FBI to tell them that I was in India to encourage the Communist takeover of the country, that I had already encouraged the Indians to take over Goa, that I had once "praised the Russian education [sic] system" in *The Saturday Evening Post,* and that I had been responsible for a visit to India by Mrs. John F. Kennedy. A broad spectrum view of subversion. He identified me as Kenneth D. Galbreath. This intervention was taken rather seriously; the Birmingham man, who claimed to have met me during the war, was accepted as an expert on my past. Thereafter, when any question arose, the FBI went back to see him.

In 1963, I returned from India and spent another few weeks in the White House. Again no one alerted the FBI, although this time there might have been reason. I had been asked by the President to represent the United States in working out the basic arrangements for a new agreement on air flights between Canada and the United States. Until then, not having anything to give in return, the Canadians had been severely restricted in their flights to Florida, California and other American centers of sunshine and rest. In a highly irregular but extremely efficient gesture, Mike Pearson, then the Canadian Prime Minister and an old friend, told Kennedy that for these very preliminary findings and recommendations, since I had often praised myself as a onetime Canadian, I could be considered as representing Canada too. So I did — a clear case of divided loyalty. Negotiating with myself, I readily reached agreement. The arrangement showed that loyalty, like being a security risk, can be an on-and-off thing.

The next burst of concern, considerable but hardly approaching that of 1961, came in the autumn of 1964. Lyndon Johnson appointed me that year to a board that was to oversee the poverty program, something which I had had a small hand in developing. Though I had been an ambassador, no risk could be run; association with the poor, far more than with diplomacy, has always brought out the strongest in left-wing tendencies. The files were searched and the field offices put to work once more, although now with a certain delicacy and restraint: "assign to experienced personnel and conduct no neighborhood investigation unless some reason for doing so arises, at which time Bureau approval should be secured." The principal new discoveries were that I had served as ambassador and also as a consultant on the "President's Commission on Heart Diseases, Cancer and Stroke (no dates indicated)." Of my onetime Commerce nonemployment I had eventually been apprised; of my unservice on heart, cancer and stroke I had never heard until I got the file. The FBI also learned that "the appointee appeared to take pleasure in criticising the Department of State and its policies, while serving the Department as our Ambassador in India," and the man in Birmingham was visited by an agent in what the files call a Special Inquiry. He now conceded, rather handsomely, that "he had no specific information that GALBRAITH was a Communist or enemy agent." He did think it significant that in a photograph taken at the time of Nikita Khrushchev's first visit to the United States I was shown standing next to him and he "suggested KHRUSHCHEV may have requested GALBRAITH's presence." Once again my appointment went through. However, I was detached in an administrative shuffle when I began making speeches against the Vietnam war.

The Vietnam war produced my last important encounter with the FBI, although there were a few minor brushes unrelated to that conflict. In 1968, an internal memorandum had reviewed my novel, *The Triumph*. "The book primarily is a 'spoof' and satire against the State Department, Dean Rusk and American policy to uphold dictators in power for the reason of overthrowing communism . . . Several miscellaneous references are made to the FBI, but nothing of any pertinence. The references are not derogatory." The following year two commencement addresses were given in New England which the Boston SAC (Special Agent in Charge) thought worthy of mention. One speaker attacked me, praised Hoover and said in a further letter to the *Boston Globe* that "No student of Hoover's ever burned his country's flag, beat up his instructors, or screeched obscenities at school the day he graduated." The other speech was mine, criticizing Hoover. The agent thought Hoover might want to send a letter of thanks to the first speaker and have a transcript of my speech, which the agent promised to get.

Sometime in the mid-sixties President Johnson summoned me to Washington to work on some plans having to do with food for India. I met him at the plane at Kennedy Airport — with J. Edgar Hoover he had been in New York to attend the funeral of the wife of Emanuel Celler, long the head of the House Judiciary Committee. I hadn't seen Edgar for many years; I thought he lacked affability and conveyed, in fact, a certain aspect of disapproval and mistrust. This must have deepened as the Vietnam war became a major issue.

On July 10, 1967, and again on December 6 of the same year, the White House asked the FBI for information on me and was wonderfully candid as to kind and purpose. The first asked for a name check on Galbraith (and three other individuals) "who allegedly are endeavoring to raise money for the reelection, during the coming election year, of a number of 'Dove' U.S. Senators." I had been so engaged, with much success. People who couldn't do anything else about Vietnam positively liked to give money. This highly improper request was filled, and as to the impropriety the FBI was not itself in doubt. Hoover carefully advised the White House that "A copy of this communication has not been sent to the Attorney General."

The later request in December from the White House was more specific as to what was wanted, for, in responding, the FBI said: "The following is being furnished in reply to your request for the results of any investigation conducted concerning the above individual [this being Galbraith] *wherein information of a subversive nature was developed."* My italics.

Once again nothing happened; as always in the government of the United States, evil intention is only marginally related to evil action, a fact of which all who are in any way susceptible to paranoia should be aware. The memoranda submitted were, apparently, the previous crap. But no one at the White House had any business asking such questions for such a purpose; whoever did is morally if not legally on a par with the Nixon men who went to the minimum security slammers. Nor had Hoover any business responding.

However, the Nixon men also came to offend in the same way. On October 6, 1969, around eight months after Mr. Nixon came to office, his counsel asked for information about me, none of which could have been for the purpose of offering employment. He was sent material that had gone over under the previous administration.

This White House intervention leads me to my concluding thought. Once many years ago my wife worked with the FBI as a language expert through a long trial of alleged Nazis in Newark. She was struck with the extreme decency of the individual agents and especially with their effort to establish the bias of anyone who was providing information adverse to a suspect. "We want to know their angle — what axe they have to grind." The decency is generally manifest in my file — in the faithful reporting of favorable comment and the warnings that I have noted against those with an angle. What was wrong with the FBI was the archaic, angry and, in the end, senile old despot who headed it and the people who were too frightened to retire him. Also the people, as at the White House, who used it for their own political ends. Also all who acquiesced in the scrutiny of subjective beliefs and attitudes — including those of us who responded tolerantly, as though they were needed, to questions about the loyalty of men such as Adlai Stevenson. Also, and perhaps most important, those who saw Hoover, his anti-Communism and the FBI as instruments against liberals, and the officeholders, including the liberals, who went along out of fear. It is impossible not to have fun at the expense of the FBI. But there is need to distinguish between the members of the FBI rank and file and the people who so egregiously misused them.

14

The North Dakota Plan[1]

IN TIMES PAST, I have found a certain amusement in maintaining a watching brief on international affairs and then enlarging on its inherently comic tendencies. The Dulles brothers, as an example, were a fine opportunity to this end. With very little effort one could picture John Foster, with his capacity for losing friends and making enemies, and Allen, with his talent for truly spectacular misjudgment, as agents of the Communist conspiracy which each so feared and featured. Any issue of *Foreign Affairs* provided inspiration for a dozen imagined articles, all equally unreadable and each with a slightly more ludicrous title than the last. One that came almost automatically to mind, ghostwritten for the Panamanian foreign minister, was "Panama Looks North and South." Another, an original work by one of my very reputable Harvard colleagues, would be called "Germany Divided." A third by the Finnish foreign minister would be "Finland Looks East and West."

Of late, however, I've been stopped by the Buchwald Syndrome, something I've mentioned on other occasions. This is the problem that Art Buchwald faced during the days when

1 In 1978, a Paris-based organization called APHIA (Association for the Promotion of Humor in International Affairs) made its annual award to a presumptively innovative figure in the field. In accordance with its rules, the recipient was required to make a significantly imaginative contribution to the art of foreign policy. This, as the recipient, I tried to do; thus the North Dakota Plan.

Ron Ziegler was promoting Richard Nixon as a model of rectitude and public grace. Nothing Buchwald could invent, he said, was as funny as the original; there was no room for improvement. And so it is now with international affairs.

Thus I do not believe that anyone can improve for amusement on the thought that Ethiopia is now a workers' paradise. Or that Somalia, which was last year's workers' paradise, is now a bastion of liberty and free enterprise and that soldiers fighting well inside what used to be Ethiopia were recently the victims of Ethiopian and also Soviet and Cuban imperialism.

Nor could any humorist surpass the recent efforts of my onetime Canadian compatriots to explain to an Eskimo named, I believe, John Smokehouse, that he shouldn't pick up radioactive fragments of a space satellite. John, it was explained by the distinguished engineers and scientists involved, had never heard of radiation or satellites. Nor had he ever heard of space.

The last great generation of Frenchmen and Englishmen defended their empires as the cutting edge of Western civilization. Now the adequately motivated sons and daughters of the empire-builders are defending their homelands from the people who were so civilized. Only one generation separates the two sets of heroes. God, I believe, so arranged things. He wanted some relief from the terrible solemnity of His new companions, Larry Flynt and Chuck Colson.

Only where American foreign policy is concerned can one be more lighthearted than the reality. That is because Barbara Walters and Walter Cronkite, who are taking it over, are terribly sober and responsible people, as all can see on television. So, of course, is Cyrus Vance as he looks on.

Since foreign policy offers no opportunity for noninherent humor, one is forced to be serious and constructive. In consequence, I've worked out a grave plan that will relieve the tensions of present-day international relations. No one will be in doubt as to what any such plan should do:

Great power rivalry must be eliminated; it is dangerous.

Ideological conflict must be turned into peaceful indifference.

There must be no cause for quarrels over international boundaries.

Armies and navies must be curtailed.

Political ambition must be reduced.

To the greatest extent possible, all countries must have a good ethnic mix. As President Carter once said, or greatly wished he had, there is no case for ethnic purity.

The plan I have developed accomplishes all of these things. It invites, as world government does not, the support of those who affirm that small is beautiful. The plan is associated with one of the deathless names in the field of innovative international action, a name especially evocative here in Paris, that of Bismarck — Bismarck, the capital of North Dakota, To coin an old phrase (as Samuel Goldwyn once said), I believe the hour has struck. The time has come for the North Dakota Plan.

In the North Dakota Plan the map is the message. Every needed reform in international relations can be achieved if national boundaries are simply redrawn so that all countries, without exception, are the shape and size of North Dakota.

All boundaries would then follow the lines of latitude and longitude. These are well known or can be discovered without difficulty; accordingly, there could no longer be any boundary disputes.

One exception to the strict rectangular form would be permitted. That is where, as in the case of the eastern boundary of North Dakota, the new national territory impinges on water. Then it would accept the natural boundary and stop rigidly at the high-water mark. This is vital, for it keeps the Plan from being a device by which countries can own any ocean.

Also, the eastern boundary of North Dakota is the Red River. This makes the North Dakota Plan attractive to the Soviets and also, and especially, to the Chinese.

Under the North Dakota Plan great power rivalry disappears. This, as competent logicians will agree, is the plausible consequence of there no longer being any great powers.

The arms problem is partly solved. Few countries will have a seacoast; none will have an ocean. That eliminates navies. In any case, one does not readily think of North Dakota as a naval power.

Other armaments involve more serious difficulties. But they are taken care of by what the cognoscenti call the Eugene V. Rostow Doctrine. Working with the Pentagon and the Rand Corporation and therefore getting wholly reliable advice, Rostow's Committee on the Present Danger has calculated that seven Minuteman missiles and ten Cruise missiles in each of the new states will keep every one of them completely safe from nuclear attack.

In the interests of an equal start, it will be necessary to dismantle the antiballistic missiles now installed in the actual state of North Dakota. But they were to be phased out anyway as an economy measure. Indeed, a few weeks ago legislators from North Dakota proposed, as a compromise, that the sites be manned only during working hours. This reflected the belief that the Soviet Union, the workers' fatherland, is no less firmly committed than the United States to the eight-hour day and the forty-hour week.

In many if not most of the new countries ideological differences will be irrelevant. Not many will worry over whether the several Dakotas that occupy northern Siberia, the central Sahara, most of Australia, Greenland or the Gobi Desert are capitalist or Communist, although I have to admit that a certain number of Americans may. Dean Rusk and maybe Henry Kissinger will also question the reference to the Red River.

The North Dakota Plan will reduce political ambition and associated tension. This it will accomplish in the only possible way, which is by satisfying such ambition. Any person who wishes to become a president, a prime minister or, reflecting a more affluent modern goal, a shah, an emir or a sheik, can move to one of the completely unpopulated Dakotas of the world and set himself up there as a chief of state. It is well known that many of the more enterprising breed of modern politician

can overcome the apathy they arouse in no other way. The plan
will be especially attractive to Senator Howard Baker.

The man whose ambition has been so satisfied can then go
on state visits to the heads of the populated nations, and here,
drawing on my own observations in India where there was a
state visit nearly every week, I wish to make a truly serious
point, if a slight digression. It concerns the reason state visits
are made. They are not to discuss business. That can almost
never be risked, for serious conversation between great men
can easily lead to uninformed agreement. State visits are made
because the visitor is greeted with a combination of pageantry,
food, alcohol and affection that he knows he will never receive
or deserve at home.

The boundaries established by the North Dakota Plan will
unite French and Germans, Arabs and Jews, Indians and Pakistanis, and keep the Scotch from trading the high road to
England, which Dr. Johnson rightly identified as one of the
noblest prospects ever seen by one of my race, for a mess of
North Sea oil. In all cases international tension will be turned
into harmless local hatred. But internal tensions will also be
reduced. In the separate successor nations of the former United
States there will be fewer conservatives to hate liberals, fewer
whites to fear blacks and fewer Georgians to arouse sympathy
and sustain condescension.

The key to reduced tensions, national and international, is
the borders — borders that cut straight through every animos-
ity, however cherished. These borders give us the slogan, the
letters that, inscribed on our banners, mark the end to international relations and thus to all resulting sorrow. The
letters are reminiscent, lovable and different — KIA, "Keep It
Arbitrary."

I come to one final point. The question has been asked in
the United States in these last weeks if the North Dakota Plan
will not impair national sovereignty. On this we have the assur-

ance of Ronald Reagan, former and future presidential candidate and one of my colleagues as a founding pillar of Americans
for Democratic Action. Sovereignty, Governor Reagan has
pointed out, is a good thing. The Panama Canal treaties which
he opposed were a bad thing, for they diminished the most
fragile form of sovereignty — sovereignty we never had. But
the North Dakota Plan, in contrast, enormously enhances the
number of sovereign states. There will, by my calculations, be
twenty-seven in Western Europe, one hundred and twenty
in North America, one hundred and twenty-three in the
USSR and eighty-five in Antarctica alone. Thus will the sum
total of sovereignty in the world be increased. No one can be
against that.

15

Berlin

MY FIRST VISIT to Berlin was in the summer of 1938. Although I was a liberal, a New Dealer, a confirmed anti-Nazi and otherwise possessed of all the reputable views of the time, there seemed nothing inappropriate about spending a relaxed summer in Hitler's Germany. Partly this was because there was much curiosity about German economic policy at the time, and this seemed to justify my presence. By one school of economists, Hjalmar Horace Greeley Schacht, who was assumed to be Hitler's economic *éminence grise,* was thought to have worked miracles with the German economy. By another school, he was thought to have resisted unsuccessfully the measures that would soon bring its total collapse. Both schools, it later developed, were wrong. He had done almost nothing. Hitler used Schacht to provide a veneer of respectability for economic policies — deficit spending for employment, control of incomes and prices — that were then thought insane and which other countries eventually adopted. When the war came, he was retired. Some associates of mine in military intelligence matters interviewed him in 1945. They discovered that he was very angry but deeply uninformed. This became evident at Nuremberg, and, his reputation as Hitler's banker notwithstanding, he was acquitted.

More important than the need to study the German economy was the feeling in 1938 that to be an American was to be de-

tached from responsibility. One could study wickedness and op-
pression without being in any way related to it. Earlier that
summer my wife and I had been in Vienna; this was in the days
immediately following the *Anschluss*. Then we went on to
Prague. Outwardly both cities were calm. If there was an underlying current of fear, it was something to be noted carefully
and, of course, with sympathy, and reported on when one got
home. Late one night we crossed by automobile from Czecho-
slovakia into Germany north of Karlovy Vary or, as was then
preferred, Carlsbad. On the Czech side parties of soldiers were
moving along the road, and we were stopped at checkpoints
for inspection. We had debated earlier that day whether it
might not be a good idea to buy an American flag to fly on
the front of our Ford. It would be, we thought, a badge of
innocence.

Ever since that summer, Berlin has seemed to me one of the
most interesting cities in the world. I have a simple test: it is
whether, no matter how many times I've been there, I still feel
a glow of excitement on the moment of arrival. That glow has
always been present on going into San Francisco, strangely on
landing at Dum Dum Airport in Calcutta and on arriving in
Paris. In May of 1974 — my most recent visit — a friend with
an unremitting interest in good works asked me to come with
him to Berlin for a few days; with the support of equally well-
meaning Germans and Americans he is organizing a discussion
center somewhat along the lines of the Aspen Institute in Colorado in a villa on the Schwanenwerder peninsula in the Havel.
Here, he hopes, scholars, politicians, officials and businessmen
or plant managers from Eastern and Western Europe, the
United States and the rest of the world will meet to consider
matters of common interest. It was thirty-six years since my first
visit to Berlin, and I had been back often in the interim. But
on going into Tempelhof Airport there was, once again, the
glow.

*

During the summer of 1938, we lived for a while on Pots-
damerstrasse and then moved to a pension on the Kürfürstendamm. We were only recently married, and my wife, who had
studied in Germany, had many friends there. All were dis-
creetly anti-Hitler, and it was something of a shock to find that
the aged husband of our landlady was a passionate Nazi. He
was, I believe, the only avowed Nazi we met that summer; even
the guide later provided by the German agricultural ministry
to take us on a visit to the estates of the Junkers in East Prussia
— I was then interested in agriculture, and the Germans had
rather overestimated my importance in the field — assured us
at the end of the tour that his party membership was purely
a matter of necessity. Once a classmate of my wife's, who was
doing his military service with a cavalry outfit north of Berlin,
came down to join us and, contrary to regulations, shed his
uniform for the visit. Whenever a party of soldiers came in
view, he turned and studied the nearest shop window, once
giving detailed attention to a display of bathroom fixtures. He
had chosen the cavalry because he liked to ride and thought it
would be useless if Hitler went to war. Later he served on the
Eastern Front, was taken prisoner in the last hours of the war
and survived five years in Russia, getting over one difficult
spell by volunteering for work on a dairy farm where he had
unlimited access to milk. On his return he joined the German
diplomatic service. We last saw him at the christening of one
of his children by Richard Cardinal Cushing. He was then
serving as Consul General in Boston and, like most others in
the vicinity, had become very fond of the Cardinal.

Life in Berlin in 1938 was not very expensive, and we had
enough money to make the rounds of museums, galleries, palaces and restaurants and, in the evenings, the music halls and
cabarets. Some of this was because such activity seemed compul-
sory in Berlin. To go home in the evening was somehow to
concede defeat; Dorothy Thompson and Sinclair Lewis in their
day would not have done so. One of the principal tourist centers at the time was the Haus Vaterland on Potsdamer Platz, a

kind of Disneyland of regional restaurants, cafés and cabarets. It still stands today, a dismal, blackened hulk, one of the few buildings in Berlin that has been neither rebuilt nor removed.

I was being sustained in Europe for educational purposes by the Social Science Research Council and, ultimately, the Rockefeller Foundation and was recurrently assailed by the feeling that I should justify their outlay. So I went to visit economists, some working (sometimes apologetically) for the government, some at the once notable Institute for Business Cycle Research (*Institut für Konjunkturforschung*), a few who had been rendered idle by the Nazis. One of the latter was Professor Max Sering, in his day a notable authority on agricultural economics. He had disapproved of National Socialist economic policy from the outset, mostly on conservative grounds, and soon after the advent of Hitler had published a pamphlet adumbrating his objections. It was promptly suppressed. I visited him one day in a large house on a tree-lined street in Dahlem. He reiterated his objections to Hitler in a voice that could have been heard through the open study windows for half a block. Since no one was there to listen, I judged that the Nazis didn't much care. Years later in Cambridge I had a visit from another scholar whose acquaintance I made that summer. Not long after we were in Berlin, he had been sent to Dachau by the Nazis as politically unreliable. Then during the war he was released and rehabilitated and made a regional commissar for food and agriculture. He had tried hard to prove his patriotism and after the fall of the Reich he was put back in Dachau as a war criminal. When he came to see me, he had just been rehabilitated for the second time. He was, like Schacht, a bitter man.

National Socialism was not obtrusive that summer. Uniforms were few; we never saw Hitler, Goering, Goebbels or any of the other high Nazis. Possibly we lacked curiosity. Earlier, in Munich, my wife had lived in a student hostel with Unity Mitford, Hitler's admired and admiring English friend. Miss Mitford, always after much preparation, departed at intervals for appointments with Hitler and returned to tell of them. As a result, Hitler's name was rather a household word around the *Studentenheim*. He was taken for granted by the students and by us.

Our only noteworthy encounter with the Nazis occurred one day when we had to go to the neighborhood of the Wilhelmstrasse to get the papers required for taking an automobile across the Polish Corridor to Danzig and East Prussia. The Nazis were, that day, welcoming a more-than-routine functionary of the Italian Fascist Party — its Secretary-General or some such, as I recall. The Brown Shirts (the SA) had been assigned to line the streets from the Kaiserhof Hotel to Templehof, a considerable show. Evidently they had been ordered out several hours too early, for, by the time we arrived in the neighborhood, they were visibly bored, and many were fairly drunk on beer passed out from the bars. Traffic was much disrupted, and somehow we took a wrong turn and found ourselves going the wrong way down the Wilhelmstrasse between the two disorderly lines of Storm Troopers. The top of the car was down. Seeing something to break the tedium, an American Ford at that, the officers called their men to attention, and they quickly deployed across the cross streets to keep us from turning off. Hands, some still holding beer mugs, came up in the forearm salute. The shouts of "Heil Hitler" were, if not deafening, at least uproarious. We faced the prospect of going all the way to the airport before we were released. However, after a few blocks, we came upon a streetcar which was cautiously making its way across the street. We turned in behind it and followed it off. A picture of the occasion featuring the salute, had it surfaced three or four years later in Washington, would have attracted attention. I was by then running wartime price control and was widely charged with employing authoritarian techniques.

I was next in Berlin in the summer of 1945. This was as one of the directors of a considerable organization, the United States Strategic Bombing Survey, which was assessing the effects of the air attacks on the German economy. The members of our group were clad in indifferently fitting uniforms without insignia and were regarded with suspicion by legitimate officers whose brisk style we tried, unsuccessfully, to imitate. The American forces had then just moved in, and although the city had fallen several weeks earlier, streets were still blocked by tanks and self-propelled guns, and the rubble still smelled of decaying bodies.

By this time we were no strangers to bombed-out buildings and devastated cities. But Berlin seemed the ultimate wasteland. The Tiergarten was the worst. It was filled with the detritus of battle; the distant skyline was made by the jagged, formless edges of the ruins. It was evening when I first crossed the park, and the setting sun had an orange brickish color from, so it was said, the dust of bricks and stones still being carried in the air. I remember thinking, rather imprecisely, that on the day the world ended it would look like this. The more common reference was to the landscape of the moon, but that also was imprecise. We have now seen the moon landscape, and, though austere, it is less alarming.

Russian troops were still passing through the city — dirty wagons, small shaggy horses, ragged harness, dusty men. In appearance they must have been closer to the formations of Genghis Khan than to the American motorized battalions that were also moving in. In front of the new Reich Chancellery a Russian guard was doing a stirring business in unissued medals that had been found inside. For my wife and several friends, I bought copies of the special Honor Cross that Hitler had issued to exceptionally fertile mothers and an Iron Cross for Harry Luce, from whose employ I was then on leave.

The vast Chancellery, the work of Albert Speer, was a partial but by no means complete ruin. Many of the great ceremonial rooms were nearly intact. The furniture was still in place but stripped of its leather by Russian soldiers much in need of it, I believe, for footwear. A few weeks earlier, under examination in Flensburg, Speer had told us at length of the climactic hours in the bunker below the adjacent garden — a story that has been retold a hundred times since and, as compared with the first telling, with astonishingly little variation or improvement. A Russian sentry, when we went to see it, was prohibiting all entry to the bunker. This whole side of the Wilhelmstrasse — which also included the Presidential Residence — is now a waste of irregular grass stretching away to the Wall.

Then, as ever since, the meeting place of East and West was a few yards away at the Brandenburg Gate. Here in the evenings American and Russian soldiers met, almost exclusively to do business. Watches were, of course, the main item of commerce. One night I went there with Paul Baran, later to become, with Paul Sweezy, one of the two most respected Marxian economists in the United States — and as a professor at Stanford almost the only Marxist to hold a senior academic post. Baran was then my immediate assistant and possibly the most remarkable technical sergeant in the United States Army. His father, before the revolution, had been in exile in Switzerland. While there, his interest had shifted from politics to medicine, and he had emerged, after World War I, as one of Europe's leading specialists in lung pathology. After teaching for a period in Berlin, he returned in the mid-twenties to the Soviet Union. There, some time after Stalin came to power, he received word from an old friend that if his son was not over the border within a day or two, he would face a long period of corrective detention, presumably in Siberia. Though a persuaded Marxist, Paul Baran had always been a man of inconvenient independence who expressed his views with a terrible combination of clarity, humor and derision. Using a Polish passport, he got away to Finland and made his way on to Berlin, where he joined the university and made his living partly writing advertising copy, partly writing theses for less qualified doctoral aspirants. One piece of his advertising copy became famous. It was for a condom, the name of which I vaguely and perhaps incorrectly remember as Nims. The ad showed a tombstone; on it were engraved the words: HERE LIES NO ONE. HIS FATHER USED NIMS. Later in 1945, Paul and I went to lunch one day outside Tokyo with a Japanese businessman. While representing one of the big Japanese combines in Berlin before the war, he had gone to the university, and Paul had written his Ph.D. thesis for him, a distinguished piece of work on Japan's future in the world economy. It had later been published in Japan, and Baran asserted that it had won the nonauthor an honorary professorship in a major Japanese university.

Although he would dearly have loved to become an officer, Baran never made it. This was partly owing to his appearance; his stomach bulged over his web belt, a tendency that was only slightly disguised by his shirt, which was often outside his trousers. Once at our headquarters in Bad Nauheim he showed up on parade in carpet slippers. He couldn't remember to salute except when it took officers by surprise. They often assumed from Baran's face that he was laughing at them, an impression that was accurate. Some weeks after the surrender he uncovered, living in Wiesbaden, Franz Halder, former Chief of the German Army General Staff, who was fired by Hitler in 1942. Halder had expected to be flown to see Eisenhower on the day the Americans arrived. Instead he had been left in solitary neglect. For reasons that had little to do with our task but much to do with Baran's curiosity concerning matters which had earlier been his responsibility at OSS, he interviewed Halder at great length. When he was finished, the general asked if the American Army had many intelligence officers of like quality. "If it has," I remember the transcript reading, "it explains much about this war. I may tell you that your knowledge of the problems facing the Wehrmacht on the Eastern Front is markedly greater than was that of the Führer."

At the Brandenburg Gate the night of our visit Paul questioned a Russian officer on the allegations, already common from the German population, that the Russian soldiers were being very rough with women and given to rape. "We have walked all the way from Stalingrad," the officer replied, gesturing toward a group of German girls who were loitering ostentatiously nearby (and who, it should be added in compassion, were probably hungry), "and we are not inclined to take unnecessary exercise." A day or so later Paul told me that, again for reasons of his own, he had been to interview the head of the largest construction firm in Berlin. He had found him in a large office on top of one of the few undamaged office buildings in the city. The windows looked out on limitless acres of devastation. The builder, a man of advanced years, gazed contentedly on this wasteland and told Baran that he thought the building industry faced a *Hochkonjunktur*.

With George Ball, later Under Secretary of State and Ambassador to the United Nations and then my colleague in the investigation of the effects of the air war, I attended, as an uninvited volunteer, the Potsdam Conference. Truman, Stalin, Churchill and later Clement Attlee. I've previously told the story[1] and will spare myself the pleasure of repetition. Despite our presence, Potsdam seemed an inordinately amateurish effort. Certainly it lacked the grandeur one had always associated with Versailles. Also, while affirming the division of Berlin into sectors and Germany into zones, it left largely undefined the relation between the sectors and the zones, the occupying powers and the Germans. The authority of the Allied Control Council was vague, and the relation of Berlin to the rest of Germany was susceptible to differing interpretations. The question of reparations, including amounts and whether they were to be from existing capital equipment or renewed production, though discussed at length, remained obscure.

Here were the sources of much future conflict. Instead the Potsdam Conference was the prelude to a third of a century of peace (as this goes to press), with the prospect of more. There also ensued the greatest prosperity that Germany and Europe

---

[1] In *The Age of Uncertainty* (Boston: Houghton Mifflin, 1977), p. 229.

ever experienced. Maybe the sheer fragility of the agreement caused all concerned to treat it with more caution than would otherwise have been the case.

Berlin was much on my mind the rest of the summer. When it had fallen, nearly all of the high Nazis escaped to the West, in accordance with the sound estimate that the Americans and the British would run more agreeable jails than the Soviets. One exception was a man we much wanted to see — the chief statistician of the Speer ministry, whose name was Rolf Wagenfuehr. In the closing days of the siege of Berlin he had expressed interest in what the Soviets would be like; some of his colleagues whom we had earlier arrested said that maybe he had been a roast-beef Nazi — brown outside, red inside. Wagenfuehr was needed for interpreting and advising on the reliability of the mountain of statistical information which, by this time, was in our hands. Speer, among others, had stressed his importance. Presently we heard that he was living in West Berlin, and I sent Paul Baran there to bring him to Bad Nauheim.

In those days, like other intelligence agencies, we had the right of summary arrest. However, except in the case of military men and the most senior and criminal of the Nazis, the power seemed unnatural, as indeed it was, and I had instructed that we use it with restraint. Lesser civilians were to be given a day or two of notice. Given such warning, Wagenfuehr removed himself and his wife to East Berlin, where he had already found employment putting the statistical services of the Reich back together for the Soviets. Baran recruited a soldier or two, went to East Berlin, found him in bed and brought him out of the city. The Soviet occupation authorities, not surprisingly, were outraged, and not long after Wagenfuehr arrived in our small detention center, word came, I'm not sure how, that a severe protest was making its way from Marshal Zhukov to Eisenhower. In those days Ike was thought to have a bad temper. Superiors were held responsible for the actions of their subordinates, and I was the superior. Baran was also having second thoughts about his own exploit. His association with the action had become known; Stalin was then alive; and Baran's father and mother were still living in Moscow.

Another onetime Berliner came to our rescue. This was Jurgen Kuczynski, a devout Communist who was working for me on the problem of manpower supply in the wartime Reich. Kuczynski had been petitioning at weekly intervals to go to Berlin where, among other things, he wanted to see what had happened to his library. He now assured me that such was his reputation with the Soviets that he could take Wagenfuehr back, reinstall him in the *Statistisches Amt* and otherwise calm Soviet tempers if only I would authorize his trip. He was as good as his word. A Boston lawyer, James Barr Ames, then a colonel with the Survey, went along at my request to supervise what amounted to the unkidnapping of Wagenfuehr. Jim accompanied Kuczynski and Wagenfuehr to the Soviet sector where Kuczynski was very warmly received by his comrades. They did not press their complaint.

During the week or two he was in custody, Wagenfuehr worked diligently to advise us on the scope and reliability of the statistics with which he had been concerned. Later he moved west to assume a research position with the German trade unions and once asked me to come back to speak at a big trade union conference. I was unable to go. In the latter part of 1945, Kuczynski returned to Germany and to East Berlin to occupy various senior political and (later) academic positions in the GDR. A few years ago I had a letter from him asking me to ask the State Department to reverse a decision that had denied him a visa to lecture in the United States. He didn't mention having saved me from the Eisenhower wrath. I tried but failed. The grounds for refusing the visa — for not waiving the bar against Communists — were, as usual, unworthy.

During 1946, I was again in Berlin. Earlier that year William L. Clayton, then Assistant Secretary of State, asked me to take charge, in a manner of speaking, of economic affairs in the occupied countries — Germany, Japan, Austria and Korea. We were both impressed by the knowledge of these countries I had accumulated while investigating the effects of the bombing. As I've told elsewhere in these essays, it was a highly unsatisfactory job. Neither General Clay in Germany nor General MacArthur in Tokyo considered himself in my line of command. Occasionally, in those months, we amused ourselves in the Department by imagining that MacArthur might declare total independence from the United States — UDI as it came to be known with Ian Smith and Rhodesia. What would we do? There was similar uncertainty at all levels at the time as to what our policy in Germany was or should be. Some thought we should foster recovery and encourage rehabilitation of the German industrial plant. Otherwise the Germans would be a heavy burden on our backs. I leaned in this direction. And an economist could not look at a demoralized and prostrate economy without contemplating the steps that might revive it. But the Morgenthau Plan for reducing Germany to a pastoral and petty bourgeois state still had vigorous adherents. They warned with no little vehemence of the dangers of a resurgent Reich. To restore Germany was also to risk arousing the Russians. One who felt this danger keenly, and I thought properly, was my close associate, Walt W. Rostow.

By the summer of 1946, Americans, officers and civilians, had settled in nicely in Berlin — many thought too nicely. They occupied great shaded mansions in Dahlem, Zehlendorf or out in Wannsee. House servants, gardeners as well as drivers had become an accepted part of the American way of life. Cooks were compared; the eccentricities of the German servant class were the subject of much animated conversation. The second in command in the State Department mission under Robert Murphy, an old friend from my agricultural days, concerned himself exclusively with his housekeeping. But in other houses more of the conversation, which was long and intense, was about the occupation, that is to say what to do about the Germans and the Russians. The Nuremberg trials were then in progress, and denazification was a question of continuing, although receding, interest. More attention was being given to affirmative programs, as they were called, notably re-education. Hitler, it was being observed, was not an accident; he was the natural product of an authoritarian family and social tradition and an even more authoritarian educational system. Intelligent remedial action required that these be democratized, the previous damage undone. How this was to be accomplished was not clear. It was not yet realized that although the Germans were the captives of their educational tradition, Americans were the captives of their social science.

A more practical discussion was on coal and food, two matters that were closely related. Coal was desperately scarce and remained so, partly because the men who mined it didn't have enough to eat. The British, who had fought the war on the principle of equal rations for all, canteen feeding apart, were reluctant to give the miners extra food for their heavy work. Coal output lagged, and, in consequence, so did everything else. The problem was partly solved by giving the miners a substantial on-shift serving of thick soup or stew in addition to their ration. Since they couldn't bring containers into the pits, they couldn't take this semiliquid food home to their wives and children. To this day I can't think of this solution, which I favored, without pain.

We were, that summer, scraping the very bottom of the bins of food. The shortage was far more severe than at any time during the war. On occasion we had to ship corn to Germany to stretch out the supply of bread grain. A few months later the German Director of Economics for the British and American Zones, Johannes Semler, made a speech in Erlangen in which he voiced a number of complaints about occupation policy. In passing, he referred to the *Hühnerfutter* that we were sending to Germany. *Hühnerfutter* meant feed grain, but in military government translation it came out more literally as chicken feed. "What the Americans are sending us is chicken feed." General Clay was furious. He fired Semler and ordered a search for a more tractable, amiable and cooperative man with no Nazi taint to take his place. One was found and thus began the career of Ludwig Erhard.

Most of all, the discussion during the summer of 1946 was of the Russians. Relations with the Soviet occupation authorities were still ostentatiously amiable, and a conscious effort was made to draw them into any important social activities. I remember a large dinner party one night in a huge house on the Schwanenwerder that had belonged (I believe) to Joseph Goebbels. My partner was the handsome and, considering the time, exceptionally well-dressed wife of a senior Soviet official. She was stunned to discover that I had a passing familiarity with the writings of Marx and Lenin; she had not supposed they were much read in the American occupation establishment, and she was right. In resourceful and idiomatic English she proceeded to give me her view of the application of Marxist-Leninist principles to Soviet occupation policy. I thought it exceedingly interesting, revealing and rather logical, given the premises, and it told something of Soviet policy in Germany. Included was expression of the most profound abhorrence of anything that might risk more war. However, when I repeated it to State Department colleagues the next morning, I found it dismissed as dinner-table talk. It was a lesson I was to rediscover in later years. Professional diplomats set great store by entertainment and justify it for the important information so obtained. But then, often no doubt rightly, they discount, more or less totally, the value of all such information as gossip.

It was fourteen years before I was back in Berlin. In 1948 and 1949, the Allied Control Council broke down, and the pretense of four-power administration came to an end. The Soviets then blocked the roads, and the airlift followed. This episode, to me at least, remains something of a puzzle; I've always felt that

there was substance to General Clay's contention that had a
truck convoy pressed its way to the barriers, it would have been
allowed through. The Soviets would not have wanted to fight.
As on other occasions, having airplanes, we rejoiced in showing
what they could do. It is basic to all American military doctrine
that because we have air power, air power is decisive. But I do
not know — and one is well advised to avoid the commonest of
all errors where foreign policy is concerned, which is to compensate for inherent uncertainty with excessive certainty of
statement.

Following the lifting of the Berlin blockade in 1949, I was
in Germany again as a member of a joint German-American
commission which was to settle the question of the German
refugees from the East — those from Eastern Germany, Poland,
the Sudetenland and from the ethnic German communities,
many of long standing, that had been scattered elsewhere over
Eastern Europe. But Berlin had no major refugee problem, and
we didn't go there. Our commission had barely finished its
work before another migrant wave — refugees from the poor
farm villages of Italy, Spain, Turkey and, most notably, Communist Yugoslavia — was set deliberately in motion by the
Federal Republic. These were the *Gastarbeiter,* the foreign
workers, about which I will say more later.

In cconomics, timing is often everything. In the fourteen
years between 1946 and 1960, there occurred the West German
economic miracle. This, by June of 1960, was evident in West
Berlin, as throughout the Federal Republic. Most, although
not quite all, of the broken buildings had been repaired. The
streets were full of cars, the shops with merchandise, and the
sidewalks, day and night, were crowded with pedestrians. Kem-
pinski's, once the most famous restaurant in Berlin and one
which I particularly remembered from the days when one's
consumption of Rhine wine and Moselle was restricted only by
money, was now a hotel on the Kürfürstendamm. Not far distant was a new Hilton hotel which Arthur Schlesinger, who

was with me on this visit, heard someone refer to as the Adolf
Hilton.

My 1960 trip was to attend a week-long conference under the
auspices of the Congress for Cultural Freedom, which the *New
York Times,* in its dispatches on the sessions, described as "an
independent [sic] worldwide organization established ... to defend intellectual liberties." About 200 attended from some fifty
countries, including George Kennan, Julian Huxley, Ignazio
Silone, Stephen Spender and Constantine Fitz Gibbon. The
meetings were held in the sparkling and spacious new American-
built Kongresshalle between the Tiergarten and the Spree,
which one entered through banks of flowers, but, in contrast
with the surroundings, the world outlook was deemed, almost
universally, to be dark. The tone was set by Robert Oppen-
heimer in the opening address; he thought war probable and,
in a much quoted phrase, warned that "none of us can count
on having enough living to bury our dead." The only hope he
saw was in the resulting "grim and ironic community of interest, not only among friends but among friends and enemies."
Others were even less optimistic. Nothing, to be sure, has yet
happened to prove them wrong.

However, I was almost as troubled by the length and vehemence of the speeches as by their tone of despair. So were
others. One morning I deserted and, in a nostalgic mood, visited the animals in the zoo. Returning to the hotel at noon, I
encountered Huxley and asked him how things had gone. "Ah,
yes," he said, as only an Englishman can, "Carlos was the
speaker," mentioning a prominent South German professor and
politician. "He spoke for an hour and then, poor chap, he realized he hadn't said anything. So he went on for another hour
in what I would say was an unsuccessful effort to retrieve."

Another day, also outside the hotel, I encountered Charles
Wyzanski, the noted Federal District Court Judge and a Cambridge neighbor. He said: "Do you know who is paying for all
this?" I said I supposed some foundation, and he said, "You

should know it's the CIA." I got hold of the controller of the Congress, whom I had met skiing, and questioned him about where the money came from. He somewhat confirmed my suspicions because he didn't know. A controller, I thought, would surely know the source of honest money.

While my suspicions were aroused, my indignation remained at a very moderate level. I thought I wouldn't attend any more conferences without knowing who was paying. But I had decided not to attend any more conferences anyway. Later when it developed that the Congress for Cultural Freedom had, indeed, been financed by the CIA, I thought it best to moderate my outrage and help to put the Congress under the aegis of the Ford Foundation. There was a happier consequence. A year later in India I encountered more CIA cultural and political activities, all foolish. Supported by the President and Robert Kennedy and taking advantage of the weakened condition of the Agency after the Bay of Pigs, I abolished them all. I was less impressed by the duplicity than by the ease with which, as in Berlin, one could find out about them.

In 1960, in contrast with West Berlin, East Berlin was a desert. Cars were few, clothes were shabby, as were the shops. There were still ruins, and what had been rebuilt was in some respects worse, the pretentious buildings along the then Stalin Allee being the leading example. Unter den Linden, once so fine, was a very barren passage.

Since World War II, numerous countries have achieved a very high production of goods in face of seemingly unfavoring circumstances. Hong Kong and Singapore, city states with no natural resources of any kind, have had a major success. Austria, written off as a hopeless case between the wars, has been far more prosperous than under Franz Josef. Israel, Formosa, Korea and Spain have similarly made much out of unpromising prospects. Yet economic success continues to occasion surprise. Its occurrence is taken without question to suggest social, cultural and even moral superiority. That is how it was interpreted in Berlin in 1960, and not alone by the West Berliners. The people of East Berlin and East Germany enthusiastically agreed and were removing themselves to the higher culture in droves. In August of 1960, an estimated 16,000 moved into West Berlin, which they could do at the time by walking over or catching the subway. The year following came yet another Berlin crisis and the Wall. I was in New Delhi by this time, and at the peak of the crisis Prime Minister Nehru expressed doubt one day about the right of the Western powers to be in Berlin. I guessed that the result of his statement would be a mighty explosion in Washington. I took the evidence on Berlin over to Nehru who, in the kind of gesture that must have worried his officials no end, told me to go out and issue an appropriate correction to the press. This, with much pleasure, I did. The episode caused him to recall a conversation he had had with Khrushchev about East Germany and Berlin: "I told him that I didn't think much of the Ulbricht government," Nehru had said. "We don't either," said Khrushchev, who went on to tell of another problem. "One morning in that country they find that the plant manager has left for West Berlin. That wouldn't be so bad if his deputy hadn't left the day before, along with the chief engineer. Then," Khrushchev concluded, "they put a new man in and a week later he leaves too."

I didn't see the Wall myself until the late sixties when I was back in Berlin for a day to give a lecture for money to an international congress of advertising men. The vice president for social responsibility of a big New York agency presided at my session. He introduced me, explaining proudly to the assembled craftsmen from many lands that it was typically American that all kinds of views, even mine, were being heard. Afterward I took a taxi to Potsdamer Platz and climbed a platform that allowed one to look over the Wall to East Berlin. An East German guard examined me without interest. Like others, I was principally impressed by how badly the Wall was built. The concrete

blocks were ugly to begin with; that they had been piled one on
the other in great haste was evident.

The Wall is, indeed, an insulting thing. But it is also, like the
Potsdam Agreement, a monument to the errors of experts in foreign policy. All thought it the prelude to a deeper, sharper,
more dangerous conflict in and over Berlin than ever before.
There has been no further crisis since.

My 1974 journey was fourteen years after the Congress of Cultural Freedom/CIA convocation. In 1960, after leaving Berlin,
I went to Hamburg to give a lecture at the request of a friend,
Professor Karl Schiller, then of the University of Hamburg.
In 1974, I caught the plane to Berlin from Hamburg after a
visit with Schiller who, in the intervening years, had risen to
heights in German and European economic affairs and was now
seeking to resume his academic and business career. Hamburg
always had an aspect of solid wealth; it was now a place of unparalleled affluence. This was evident in the miles of streets
around and back from the Aussen-Alster, all lined with huge
villas and beautifully tended lawns, and also in the working-class
districts where the blocks of flats had an exceedingly spacious,
airy, even elegant appearance. It was also evident in the Vier
Jahreszeiten, which those who are knowledgeable in such matters praise as Europe's best hotel, and even slightly in the red-
light district of St. Pauli, through which we detoured on the
way to catch the plane at Berlin. One of the puzzling things
about prostitution and pornography is the blight they bring
where they flourish. Were they as financially enticing as they are
presumed to be, they should promote investment, increase real
estate values and improve the physical tone of the community.
St. Pauli suffers from the blight but much less than Times
Square.

As before, one came into Tempelhof over a big cemetery. The
terminal building was large and ugly and featured a great pro-
truding roof that sheltered the planes or some of them. Forty

years ago when the terminal was much smaller, it had a nice restaurant on the roof where one could sit in the evening and watch the Lufthansa flights taking off for every major city in the Reich. It must have been one of the first airports in the world where this was possible, and novelists made much use of it. Now there was no restaurant, and very soon there would be no Tempelhof either. It is no longer possible to have so large an airport in the middle of a city.

Still adjacent to the airport were the buildings where we were billeted in 1945. Then the Russians had just moved out, and they were indescribably filthy. I remember asking a laconic American soldier, a premature Cold Warrior, if the Soviets had stabled their horses there. "No," he said, "their infantry."

Like Hamburg, Berlin in 1974 had an air of urban opulence and cleanliness that was especially striking to anyone accustomed to New York. Except for a small area around Potsdamer Platz and in the adjacent onetime embassy section, the scars were all gone. The trees were thick and at that season freshly green, and they were now half-grown again over Unter den Linden. East Berlin was still not as obviously affluent as West Berlin, but there was no longer any very sharp contrast between the two. Checkpoint Charlie had been greatly reinforced to prevent anyone from crashing through with a car or truck (or for that matter a tank), but the guard who surveyed our Embassy station wagon and glanced at our passports was middle-aged, amiable and appreciably overweight. The streets in East Berlin were full of cars amidst which ours attracted no attention. Somewhere near Alexanderplatz there was a huge supermarket — it must occupy most of a city block. An acquaintance from the State Department, who was an authority on East German affairs, cautioned me that it was a display item, pretty much the only one of its kind. Still, he admitted that it was impressive. One is reassured to be reminded how closely we and the Communists now agree on our indices of achievement. The young in East Berlin were indistinguishable from the young across the Wall.

Both sexes wore the same abused jeans, had hair of the same length. In one pack of adolescents near Alexanderplatz were two who had "Dartmouth" on their T-shirts.

At various gatherings in West Berlin — a reception at the City Hall, a dinner one evening, a luncheon at the house of the American Minister — I encountered a small but not negligible cross-section of the Berlin political, academic and business establishments. Their preoccupations had become wonderfully commonplace. Ernst Reuter and Willy Brandt, in their days as mayor, were the defenders of West Berlin from the East. During a pleasant chat with Mayor Klaus Schutz and two of his immediate associates, problems of East-West relations came up only as one of them offered to arrange for me to visit Jurgen Kuczynski (for which, in the end, there was not enough time). Instead inflation, municipal administration, education, student discipline in the Free University, problems of the Turkish minority, the redevelopment of Potsdamer Platz, the best and most economical way of subsidizing the opera and some recently expressed views of Mayor Kevin White of Boston on urban planning were the matters discussed.

The migration from Turkey was now much more on the minds of Berlin officials than any movement past or in prospect from the GDR or East Berlin. There were at the time some 500,000 Turks in West Berlin, some thought more, out of a total population of 2.2 million. They lived frugal lives in closely knit communities, saving their money against the time when they could return to Anatolia and either buy land or go into business. But some were settling down in the city, and the city government was actively recruiting Turkish-language teachers and arranging instruction for Turkish children.

In all industrial countries there is now a new subproletariat that does the menial, dirty or repetitive work that other economically more advanced toilers no longer wish or are available to do. Blacks, Mexicans and Puerto Ricans in the United States; Algerians in France; Southern Italians in Switzerland; Indians,

Pakistanis and West Indians in Britain; and Turks and Yugo-
slavs in Germany all so serve. Yet no one quite wishes to accept
this pattern as normal and certainly not as permanent. A sub-
class of hewers of wood and haulers of water — one which is
racially or ethnically distinct — is repugnant to almost everyone's democratic sense. Thus the German euphemism *Gastarbeiter* or guest worker. When I suggested one day that honesty
required that we accept this new class structure as permanent
and face up to its implications, I had the feeling that some of
those present thought I was carrying honesty too far.

As I've indicated, one luncheon during my visit was at the
residence of the American Minister, i.e., the resident State Department official, in Berlin. This was, once again, one of the big
villas on a tree-shaded street in Dahlem. One of the guests was
a senior member of the Soviet Mission. We walked out together,
and he said he always found American hospitality pleasant; the
talk was so often about practical matters. It seemed a very nice
observation. Earlier I chatted with the commanding general of
the American forces in Berlin, who was about to depart with no
great enthusiasm for a tour of duty in the Pentagon. A mild-
mannered officer, he told me that his various army duties had
kept him from spending a great deal of time on my books, although he had been attracted by one title. I asked him which.
He said it was *How to Control the Military*. The Soviets or East
Germans were not on his mind either. He very soon turned the
conversation to the problem of race relations within the American forces and to the devastation worked by the Vietnam war
on morale and organization. In earlier years, much in the manner of the East German factory described by Khrushchev, no
sooner was a unit at reasonable efficiency than its officers and
noncommissioned officers disappeared to Saigon. Things had
since become better, and on race relations he thought the corner had been turned. It was not a problem, he observed, for
which growing up in Texas naturally prepared a man.

One morning I was collected in a car by an exceedingly handsome German woman whom I had met some years earlier when she was a Pan Am stewardess. Because of her intelligence and good looks she had been assigned to a flight that was bringing Mrs. John F. Kennedy and her offspring back from a holiday in Switzerland and on which my wife and I were also returning. We drove out through Zehlendorf to the Havel to see her new place of work and passed the zone of low buildings inherited from the Luftwaffe that once had been occupied by General Clay, Robert Murphy, Joseph Dodge, William Draper and the other archons of the American occupation. Buses make the rounds of the American installations, and waiting for one were two wives, both very pregnant. "They don't have a great deal to do," my companion said.

From Berlin I went down to the Rhineland to give a lecture at the great Benedictine Monastery of Maria Laach. (It was there that Konrad Adenauer took refuge from the Nazis.) My speech was not to the monks but to a group of German business-men who, over the Feast of the Ascension, go into retreat at the monastery to talk about economics and management. I was, however, the guest of the Abbot for dinner, in whose quarters silence is maintained except for a reading by one of the members from the Bible and a Palestinian history. I then stayed over-night in the Bishop's apartment. Though many members are old, the monastery is still at full strength. I learned, however, that for running the adjacent hotel and other auxiliary enter-prises, German nuns are no longer available. Nuns must be re-cruited from Yugoslavia. Guest workers again.

16

Germany: July 20, 1944

ON JULY 20, 1944, with the Russians fewer than a hundred miles to the east and the Western Allies known by the Wehrmacht to be on the verge of a breakthrough in Normandy, a professional German Army officer of great courage and determination, Colonel Claus Schenk Graf von Stauffenberg, left his briefcase in the hut at the East Prussian headquarters (Wolfschanze) where Hitler and his generals were assembling for the noonday briefing and went out, ostensibly to make a telephone call. Minutes later a captured British-made explosive in the briefcase went off with a terrific bang, killing four officers but leaving Hitler physically more or less undamaged. It was Stauffenberg's third try.

In Berlin, meanwhile, Stauffenberg's fellow conspirators at the headquarters of the German Replacement (i.e., Reserve) Army on the Bendlerstrasse, Lieutenant-General Fritz Thiele and General Friedrich Olbricht, were waiting to set in motion the troops, seize the installations, make the arrests and dispatch the messages calling for similar action throughout the shrinking German empire which would oust the Nazis and establish military rule by the Wehrmacht. But the news they now got from East Prussia seemed confusing so they went out to lunch and did not get back until 3 P.M.

Count von Stauffenberg, not without luck and difficulty, got by the perimeter guards around Hitler's headquarters and back to Berlin in a special plane which he had standing by. On arriving in the city, he had some trouble getting a car, a defective bit of planning, but in late afternoon he arrived at the Bendler-strasse, since renamed the Stauffenbergstrasse. Here, as chief of staff of the Replacement Army, he managed for a while to make things move. Two of the major Berlin radio stations were occupied, although, unfortunately, by men who didn't know how to run them. An officer was sent to arrest Goebbels, the highest Nazi currently in town, but was talked out of the action. Telexes were dispatched to the headquarters of the *Wehrkreise* (regional defense districts) into which Germany and Austria were divided and to Brussels, Paris and Prague. These alerted the scratch forces there available to the regional military commanders and ordered the arrest of Gauleiters, top SS officers, Gestapo officials and other inimical types. Unfortunately it seemed only proper that orders overthrowing a government should go out Top Secret, although it was hardly something that could be kept quiet for very long, and this meant major coding and decoding delays and, at the points of receipt, the need to find officers of sufficient rank to read them.

It was early evening before the messages were in hand, and by then most of the headquarters had shut up shop for the day and the responsible officers were variously attending receptions, shooting dice, having a drink or en route home and unavailable. The Gauleiters had similarly bunged off. One was celebrating his tenth and manifestly last anniversary in office, another was at a funeral and Frank, the Nazi minister in Prague, was at a ceremony opening an SS training school, as was the general ordered to arrest him. German efficiency and attention to duty were everywhere at full flood. So, except in Paris, where Gestapo, SS and Party officials were taken into custody, nothing much happened.

Meanwhile Hitler's people in East Prussia, proceeding in only slightly less confusion, had come to realize that something more than an assassination attempt was involved and were getting out word that orders from Berlin should be ignored. Finally Hitler himself went on the air to prove that he was still alive, and it was all over. On the Bendlerstrasse Colonel-General Friedrich Fromm, the commander of the Replacement Army, though he had previously shown sympathy for the conspirators, had thought it wise to sit out the afternoon under detention in his office. Now with the first display of determination of the day he resumed command, convened a court martial, had the conspirators (including, of course, Stauffenberg) convicted, one gathers in a matter of minutes, and taken down to the courtyard and shot. (Colonel-General Ludwig Beck, who was scheduled to be interim head of the new government, first tried, without success, to shoot himself.) Management was still sloppy. The insurgents were buried in their uniforms with their medals. To preserve proper indignity, they had to be dug up again next day and burned.

There followed, in the ensuing weeks, a ferocious massacre of those of the conspirators who did not anticipate their own executions. Among the casualties was General Fromm who was bumped off for cowardice. Nazi justice was not always imprecise. Justice of another sort was even visited on Dr. Roland Freisler, the unspeakable Nazi judge (so-called) who dispatched the top participants. While he was engaged in handing out automatic death sentences — usually by hanging on a rope over a hook — the courthouse was brought down on his head by a bomb. The executions continued at an informal level quite literally up to the week of the surrender in 1945.

While the action of July 20 was largely confined to the Army, the conspiracy extended to a group of conservative and aristocratic civilians and on to a number of moderate (and a few romantic) socialists. There were overtures to the Communists, who had their own operation, but this association was limited by the extreme distaste of the conspirators for such people and the very great hope of many, although it dwindled as the war continued, that with Hitler out of the way the Western Allies would happily join in a march against the USSR.

*

The story just recounted was of the supreme and, in some respects, the only moment of the German resistance to Hitler. It has been told before, and another ordinary account would hardly be needed. But Professor Peter Hoffmann — he is a German by birth and early education who has studied and taught in the United States and is now a professor of German history at McGill — has made it impossible for anyone ever to deal with the subject again, although no doubt some will try.[1] He has researched the July 20 events in Berlin, East Prussia, the provinces and in Prague, Brussels, Vienna and Paris down to the last minute and sergeant. And he has gone into all of the antecedent efforts and conversations going back to 1933. *The Times Literary Supplement,* reviewing the German original, said that it was "the essential and surely final handbook" on the subject, and the words "essential," "final," "handbook," are all well chosen.

In recent times an offensively imaginative revisionism has come to suggest that Hitler was a political and military genius who, in his lofty and statesmanlike way, was only marginally aware of the butchery of the Jews and the Poles. Much of this book consists of the case which German civilians and generals made to each other for deleting Hitler. They were not in the slightest doubt as to what his brainless military megalomania was doing to Germany or what he personally was doing to the Eastern peoples and the Jews. Indeed, the details in Professor Hoffmann's book accumulate into one of the most horrifying pictures of Hitler yet. One shudders as always that such a mad criminal with such a pack of followers could get loose in a civilized country in this century.

At a less ominous level, Professor Hoffmann's account shows the extraordinary autonomy of the German Army in managing its own affairs and in keeping its own secrets. Scores, perhaps hundreds, of high officers, including a brace of field marshals,

[1] *The History of the German Resistance 1933–1945,* translated from the German by Richard Barry (Cambridge: The MIT Press, 1977).

knew of the contract being put out on the Führer. Individuals were assigned and reassigned to facilitate the operation. The reaction was slow at Wolfschanze because Hitler's people there had no hint of the subversion. All who examined German wartime management, without, I believe, any qualified exception, were struck by its unimaginative incompetence.[2] Professor Hoffmann shows that this extended even to the field where the greatest expertise was imagined, namely political repression.

At an even less ominous level, Professor Hoffmann gives a superb, often grimly funny picture of the folk habits of the German officer class as it then was and of their aristocratic civilian counterparts. One forthright thought for giving Hitler the business in 1943 was to pull a gun on him while he was lunching with officers during a visit to the Eastern Front. Field Marshal Gunther von Kluge had to be warned so that he would keep out of the line of fire. (Kluge was passively sympathetic and after July 20 committed suicide to keep off the meat hooks.) He vetoed the method, saying, "It was not seemly to shoot a man at lunch," and adding that there might be casualties among "senior officers [including himself] who would have to be there and could not be spared if the front was to be held." Partly because there was nothing else they could do, the civilian members of the conspiracy spent their time drawing up lists of future cabinet officers which later were a great gift to the Gestapo, and getting into line the less pressing details of post-Nazi government. At one session they considered policies on the multinational corporation, and, on June 21, 1943, Julius Leber, a regular socialist leader, "met Lukaschek, Husen and Yorck of the 'Kreisau Circle' in Yorck's house; they discussed the important question of church or state schools and Leber [a tolerant proletarian] accepted the right of parents to choose."

2 This has been dealt with in Burton Klein's *Germany's Economic Preparations for War* (Cambridge: Harvard University Press, 1959). Further documentation is in the several reports of the U.S. Strategic Bombing Survey 1945–1946, of which I was a director. On returning from Germany in 1945, I wrote a piece entitled "Germany Was Badly Run," which appeared in *Fortune* in December 1945.

The military discussion, almost continuous after 1933, of ways and means of eliminating Hitler was also, I would judge, mostly a substitute for action. Any excuse, ranging from an officer becoming unavailable because of a new posting, to Neville Chamberlain flying to Berchtesgaden, to the unconditional surrender demanded by the Allies, would cause a contemplated move to be postponed. Some like Fromm were caught between the pressure to do right and the equal or deeper impulse to save their own skins. But more, it is plain, were caught up in the conflict that besets all organization men — the same conflict that is faced by a Gulf Oil executive considering a political slush fund or a General Motors man seeing a clandestine engine switch or that was the private trauma of numerous State Department and Pentagon officials contending with the war in Vietnam. The difference in degree in Nazi Germany, however, was enormous. Hitler and the Nazis were a throwback to Attila (whose Stauffenberg was the equally unsuccessful Vigilas), and with no real disguising social or moral purpose. But for the officers the ancient Prussian mystique of the state combined with the powerful tradition of disciplined military service to make dissent uniquely difficult. Americans, Englishmen or Frenchmen would not more easily have resolved such a conflict. The Latin American military *golpe* is entered upon without such difficulty, for the underlying conditions are almost exactly the reverse. The state has no similar prestige; the Latin American army is not a disciplined organization but a loose association of more or less ambitious individuals. There is no personal crisis of loyalty or discipline in acting to throw out a government. It also helps, no doubt, that failure can usually be survived.

Withal, as Professor Hoffmann shows more clearly than anyone before, the July 20 effort was a near thing. Among the high Nazis the handwriting of defeat was being read. As nearly as one can ever be certain on such matters, there was no one besides Hitler who was capable of gathering authority in his hands, taking even the feeble action that would have been required to put down the revolt and showing the much greater strength that, in face of certain defeat, would have been required to carry on. In Berlin, when threatened with arrest, even Goebbels fingered his cyanide pills. In Paris, where the military end was most clearly in view, the SS leaders and Party officials surrendered readily to the Army and cooperated as far as possible afterward in keeping word of what had happened from Berlin.

Count von Stauffenberg, as the result of wounds received in North Africa, was without one hand and a couple of fingers on the other. While getting the explosive organized in his briefcase a few minutes before the explosion, he was interrupted and had to sacrifice half the charge. (He managed to throw the unused explosive out of the car on the way to the plane.) Had the explosion been twice as strong, it is not credible that even Hitler would have expected to survive, and without his frenetic desperation, National Socialism would have come to an end that day. There would have been a different set of German leaders to reckon with in the years following, and the relations between the West and the Soviet Union would also have been interesting. Assuming, as I would, that the British and Americans rejected the overtures that would surely have come, the Western Front would hardly have been held by the Germans with the same determination as that in the east. Certainly there would have been no winter offensive in the Ardennes if the war had lasted that long. But this is not a line of speculation that is usefully pursued. No one can now tell how things would have been changed and by how much.

I first heard of the July 20 affair, newspaper accounts and speculation around military headquarters apart, from Albert Speer in May 1945. He was then a minister in the government of Admiral Doenitz, which, in a manner of speaking, was still functioning in Flensburg on the Danish border, and we were interrogating him on the effect of the air attacks on arms pro-

duction and related matters having to do with the German war
economy. He spoke of the participants in the plot — my memory is not completely firm on the point — as conservative, pa-
rochial and without much mass appeal. I remember more clearly
his criticism of their penchant for lists, for his name had appeared on one, and for some weeks, in consequence, he had been
regarded in a thoughtful way by the Gestapo. On one of the
lists in Professor Hoffmann's book Speer continues in charge of
armaments, although it is noted that his agreement has yet to
be obtained. That anyone so close to Hitler, both personally and
officially, would be thought an acceptable figure in a post-Hitler
government is a further indication of the ambiguity of the
enterprise.

During the course of that summer it became a joke among
Allied personnel in Germany that July 20 must have been the
largest conspiracy in history, for it embraced the entire German
population. Scarcely a general came into our hands who had
not, by his voluble account, been deeply involved. However,
Professor Hoffmann shows that, at the level of conversation, a
great many had been — we were right only in our guess that
the operational importance of most people's participation increased greatly after the surrender.

There was also in those days a parallel desire to detach from
the ensuing slaughter. In June or early July we were interrogating Goering, Ribbentrop, Funk, Ley and the top Nazi generals at the special high-level jail called Ashcan which had been
established in Luxembourg. Because the war crimes investigators — the Donovan Committee — had no similar personnel
and access, we had been asked to get several of the people we
were questioning on the record as to their more odious achievements. Field Marshal Keitel had presided over the Wehrmacht
honor court that turned the conspiring generals over to Freisler
for execution. He was asked how many had been so consigned.
A man of comfortably obtuse manner, somewhat resembling a
terminal career case as vice president of Chase Manhattan, he

replied that there had been none. He was then reminded of the condign punishment, possibly enhanced for the occasion, that, according to regulations, awaited those who supplied false information to the occupation forces. He asked for a chance to think, for him a time-consuming exercise, and came up with a revised estimate of a dozen or two. Someone on our team expressed astonishment at the humanity and restraint that he had shown, and later that evening he approached us on our way from another interrogation, fingering a piece of paper. There had been a further upward revision, this time to several score. The Field Marshal too had been caught in a conflict — this between the consequences of lying and the consequences of truth. Hoffmann puts the number of executions resulting directly from the July 20 events at around 200.

My final memory of these matters in that electric summer was of returning to our headquarters at Bad Nauheim near Frankfurt from a trip (as I recall) to Hamburg. It was a few days before the British election which was to choose between Attlee and Churchill, and, on my arrival, George Ball, one of my fellow directors of the U.S. Strategic Bombing Survey, told me with much delight that an unresolvable choice between military responsibility and political faith awaited me.

Earlier that day, Nicholas Kaldor, the noted economist, now Lord Kaldor and then a civilian recruit to our staff, together with Kurt Martin, another distinguished political economist who shared Kaldor's strong social democratic views, had interrogated Colonel-General Franz Halder. Halder had told them in rich and unduly firm detail of the plans of the Army to take over the Reich in September 1938 to forestall war over the Sudetenland. All the generals were set for the action when, on September 15, news came that Neville Chamberlain was flying to Berchtesgaden. This meant that the British were caving in. Kaldor and Martin wanted orders allowing them to fly to London by military courier plane — the only transportation available — to give the facts to Clement Attlee for an election speech.

Tory appeasement had been the cause of a wholly unnecessary war. With that news Attlee would be sure to win. I went to face Nicky; our meeting lasted for two or three hours.

I too wanted to see Attlee win. But if I allowed American Army transport to be used for so flagrant a political purpose, the blame and punishment would fall on me and not Kaldor. I also doubted that Attlee would use the information. Churchill had been anything but an appeaser; mention of the plot would have brought from him a terrifying rebuttal. "And to whom are my opponents now turning for support? They are turning to the defeated Naaa-zzi generals."

But my real commitment, also, was to the organization ethic. When I confessed this to Kaldor, he was distraught and deeply disappointed by my lack of character. However, a lifetime of enmity and recrimination was avoided when, a few days later, Attlee won anyway. Kaldor and I have been close friends ever since. Professor Hoffmann's conclusions make it clear that my supine course was really an act of democratic virtue. For if Kaldor had gone and if Attlee had used the information, it would have been a terrible fraud on the British electorate. Halder and Field Marshal Walther von Brauchitsch, the Army Commander-in-Chief whose participation in the takeover was also needed, were, as Professor Hoffmann shows, the ultimate in ambiguity and reluctance. Had it not been the trip to Berchtesgaden, they would, it is reasonably certain, have found some other excuse for bugging out.

17

The Indian-Pacific Train

THE MOST BEAUTIFUL railway terminal in Sydney, Australia, perhaps anywhere, is a Gothic-arched sandstone monument called the Old Mortuary Station. In civilized times the dead and the mourners came together here after the funeral, were loaded on a funeral train and, in decent dignity, were conveyed to the Rockwood Receiving House at the suburban Rockwood Cemetery. It made an interesting day; for the railway it was also, the *Sydney Telegraph* said some time ago in a nostalgic article, a lucrative run. Motor hearses eventually did it in. The final moment of the old Mortuary Station came in World War II when, somewhat tactlessly, it was used for the dispatch of troops on their way to combat in the South Pacific.

To go across Australia from Sydney to Perth, one of the few long and luxurious railway voyages in the world, one leaves from the more conventional main station which rather resembles a library and has, as its most serious pretense to grandeur, a very decent imitation of Big Ben. All American railway stations now have the empty and archaic aspect of the Baths of Caracalla. The Sydney station, especially as the 3:15 P.M. departure hour of the Indian-Pacific train approaches, has a rewarding bustle of passengers, porters, train crews and people speeding their parting guests, spouses or offspring on their way. The train itself, immensely long, is no Amtrak makeshift but a thing of gleaming silver. However, the everyday coaches and sleeping cars on the adjacent tracks remain a dull and shabby red with every indication of squalor within. The Australians have done great things for their transcontinental travelers, but one senses adherence to the accepted view that those going short distances should be subject to normal abuse.

Travel on the Indian-Pacific train — named, of course, for the two oceans it joins — is greatly approved by Australians. Just prior to its departure, my wife and I were entertained to lunch at one of the big banks by a fair cross-section of the Australian Establishment. All present were enthusiastic about our good judgment in taking the time to see their vast country by train. Our host then asked around the table as to how many had taken the trip. None had.

All should have done so. The two and a half days and three nights so spent are excellent, even memorable. The beds are long and comfortable and better perhaps than the roadbed. Each cabin, as it is called, has closets, toilet, basin and shower. Meals are by successive sittings, efficiently served and highly palatable with no false claim to elegance. The lounge cars feature comfort, the beer by which the whole Australian nation is irrigated and a piano which, on our train, was the center of an enduring amateur hour.

At exactly 3:15 P.M., the stationmaster (who, along with the head conductor, our car conductor Mr. King and the dining-car steward, had dropped by to welcome us aboard) unrolled first a white flag, then a green flag, and we were on our way. Ahead of us lay the endless stretches of the Australian outback and desert and the almost equally endless stretches of the Sydney suburbs. An hour later we were still passing square bungalows of wood or brick veneer with red tile or red-painted metal roofs, each house surrounded by a pleasant garden. The Australians were wise to choose such a large country, for of all the people in the world they clearly require the most space. Because they need so much and all wish to live within reach of Sydney or Melbourne, prices of building lots are heroic. Demand for land has somehow outrun an inexhaustible supply.

Eventually the city dwindles away, and the train runs into the mountains — the Great Dividing Range. The mountains are not very great, as even Australians concede; they lack, comparatively speaking, the wild grandeur of the Poconos or the lower foothills of the Berkshires. At precisely eighty-six miles from Sydney, the train reaches the summit and begins its descent. This also is not a moment of major excitement; the maximum elevation is around 3500 feet. We supped and then slept.

Next morning at our request Mr. King brought us breakfast in our cabin and told us in a concerned tone that, when the train changed management after Port Pirie, this grace might require renegotiation. Australia has had an interesting experience with its railways. They were built and are still mostly operated by the states. Other subordinate communities within national states have expressed their individuality in their language, music, dress or orgiastic sexual rites; the Australians expressed theirs in the width of their railway gauges. Each state had a preference; the state of South Australia had three. This is a matter where small, even marginal differences are exceptionally significant, so, when changing states, one once had to change trains. Only in 1958 did a commission — the Wentworth Committee — devise a scheme for partial standardization, and only in 1970 did through passenger service from Sydney on the East Coast to Perth on the West become possible. The Indian-Pacific train is the joint enterprise of New South Wales, South Australia, Western Australia and the Commonwealth of Australia, functioning through a joint agency called the Railways of Australia. Lindbergh, on being felicitated for crossing the Atlantic all alone, is held to have said it would have been more remarkable had he done it with a committee. Our movement is, clearly, the supreme accomplishment of a committee.

The first morning finds one still much closer to Sydney than even to the center of the continent. The country is green to

grey-green at this spring season with high grass, low fibrous veg-
etation and scattered trees — a parklike aspect. Here and there
are stretches of brilliant red or blue flowers. At wide intervals
are grazing sheep or, more occasionally, cattle. There are also
emus — vast birds on the general scale of an ostrich — and, in
something less than abundance, kangaroos. Of these one has
only a glimpse, for, as the train moves to the west, they, for reasons that are unclear, move rapidly east.

The kangaroo strikes one as an exceptional achievement in
animal design. It has two main back legs, all that are required
for efficient movement. But two supplementary front legs remain available for resting and for holding food. The pouch is
a major added convenience, and even the tail has some inex-
plicable balancing function. With everything else, kangaroos are
said to be moderately intelligent, with a slightly mean personality that earns them respect.

Returning to the countryside, it is pleasant, even beautiful,
though with a hint of more parched regions to come. One begins also to develop the feeling that there may be more of it
than is really necessary. A rewarding aspect is that one can sleep
any time for an hour, certain in the knowledge that nothing will
be missed. The farms are mercifully sparse and so are the villages — mercifully because the domestic architecture of the
rural Australians is untidy, and the landscape features an abundance of abandoned automobiles.

At around nine o'clock in the morning extreme air pollution
signals an approach to civilization. It is the great mining town
of Broken Hill. As the train stopped, I was called to the signal
office for a telephone call and was reminded how wonderfully
a train insulates you from the world. A Melbourne newspaper
wanted a comment from a visiting sage on the Australian stock
market which had crashed ignominiously the day before — the
worst slump in eight years. I was able to say quite honestly that
I hadn't heard.

The promise of aridity is postponed. Suddenly well to the west of Broken Hill the train climbs to a high plain and emerges on a land of emerald green — green pastures, green wheat fields, green oats, all running away to low, rounded, greenish hills. The farmsteads, though still sparse, are larger and range from habitable to mildly grand. The towns show pride, and grain elevators on the scale of those in North Dakota or Saskatchewan tower over all. The sheep are now dense on the pasture and, it being the season, are freshly shorn. Everywhere there is a brilliant blue flower which inquiry identifies as Salvation Jane, a most noxious weed which sheep will eat only as a last resort. At an improbable siding, our train meets the one that will reach Sydney tomorrow. In a matter of three or four minutes our crew is replaced by the men from the eastbound who are to take us to Port Pirie an hour or two ahead. A tiny town we pass has a pleasant park kept neat by a large flock of sheep; nearby is a sizable racecourse.

Port Pirie of 17,000 in population is near the head of a long, narrow bay jutting north from Adelaide. (The train runs well inland from the southern coast of the continent.) Here we paused for an hour or two. The mayor, as he occasionally does, came down to welcome the train and took us on a brief tour of the town — docks, adjacent lead smelter, parks, playgrounds, schools, new sewers, a Boy Scout camp. Port Pirie is an industrial opportunity that in this day could be unique. It warmly welcomes industry, especially American industry, wants badly to grow, is wholly unworried about air pollution and has free land for new enterprise. There is no danger that any of these attitudes will change with a leftward swing in government. Labour, which is as far left as you can go, is already solidly in. Any firm worthy of Port Pirie should get there before the rush.

At Port Pirie, or rather its neighbor Port Augusta to the north, the serious business of crossing the continent begins — the railroad puts itself on a straight course and keeps it. On one notable stretch of 297 miles the tracks do not turn whatever or

at all. The train begins on the Longest Straight at dawn of the
second morning. The first turn in the track comes at 12:35 P.M.

This is the Nullarbor (meaning no trees) Plain. It is utterly
flat, utterly empty and astonishingly beautiful. This year there
has been rain, so, most exceptionally, it is covered with green
grass and shrubs to a depth of ten inches or a foot, the shrubs
moving like waves in the constant wind. And interspersed
throughout are the flowers — white, yellow and startling red,
the last being wild hops. Along with the grass and shrubs are
patches of bright red earth, and in less favored years and seasons the voyager would see a lot more of it. On arising the second morning, we were told by our car conductor to have an eye
out for kangaroos, foxes and dingoes (wild dogs), all of which
are present and even abundant. However abundant, they were
also invisible. So now, because of the scarcity of water, were the
sheep. At Cook, a railway operating center of some importance
with a population of (by rough estimate) 200, there is an am-
biguous sign in front of the hospital, "Hospital needs your
help. Please get sick." West of Cook the hamlets — inhabited
mostly by railroad workers — immortalize Australian politicians. No town is so insignificant that it does not rate a prime
minister or vice versa. Also celebrated are Field Marshal Lord
Kitchener of Khartoum and Field Marshal Sir Douglas Haig.
Their settlements appear to be especially insignificant, reflecting, one imagines, the unquestioned distinction of the two gentlemen as the most disastrous generals in British imperial history
and the special authors in World War I of the slaughter of some
tens of thousands of young Australians.

At mid-morning, privilege raised its lovely head, and we were
invited to ride for the next hundred miles or so on the locomotive. The track stretches ahead of the diesel to the horizon, giving, quite erroneously, the impression of a downslope beyond.
The telegraph poles on their equally endless march are more
honest. They curve gently, reflecting the tendency of the earth.
In all normal experience the men who run trains are old. This

train was competently managed by an assistant driver (i.e., assistant engineer) in his late twenties and his companion a few years older. I asked Mr. Fraser, the engineer, if his run over the Longest Straight didn't cost him practice in steering. He said the problem was not as serious as the layman might imagine. Boredom is a greater threat. Incidentally, the old dead man's grip or foot pedal which, unless grasped or pressed, will stop the train has gone. Instead there is a clock which, if not punched every ninety seconds, first blows a whistle and then, if unattended, brakes the train to a halt. As Mr. Fraser and I talked, it occasionally whistled.

The assistant driver expressed a compassionate concern over the way the train blew young birds under its wheels. We passed a work crew, young and robust, standing by the track. The nearest towns to them were Port Pirie 700 miles east, Kalgoorlie 400 west. "They have," the assistant driver said, "a very quiet life." In the days of steam a third or more of all the cargo carried on the trains was coal and water to propel them. Diesels were a major breakthrough. But tank cars still go down the line two or three times a week to water the way stations. Our sojourn on the locomotive ended when the train stopped to oblige a wayside dweller who needed a prescription filled.

It must have been a fairly esoteric drug. Beside each of the tiny towns an airfield is marked out for the use of Australia's famous flying doctors. And the hamlets and remote farms or stations are stocked with drugs, each bearing a number, which the doctor prescribes by radio or land line. "One of number 16 after each meal and at bedtime. Report on bowel movements in the morning." Later in Perth a woman who had lived on one of the distant stations told me of her pleasure, even excitement, in listening in on these sessions each day. "It isn't the thing, really, but everyone does it." There was also, she said, a legitimate gossip hour when everyone within a radius of forty or fifty miles got together on the radio to exchange personal news and beliefs.

Eventually the Longest Straight gives way to some gentle

curves which then usher in another straight stretch of only forty-odd miles.

We had, by now, discovered our fellow passengers. They filled the train to capacity, and all were traveling for pleasure. Thirty or forty were members of a women's lodge in Sydney, the Sisters of Gomorrah or some such. "We are like your Benevolent and Protective Order of the Bison in America," one explained. Numerous of the Sisters had brought their husbands, and most of the women had prodigious appetites. One adjacent to us had for breakfast orange juice, corn flakes, eggs, bacon, lamb chops, toast and coffee. Another in a black silk pant suit I met in the train corridor. She backed hard against the wall. I tried to squeeze by. Eventually I made it, but there was a terrible rending sound. She said, "It's nothing." I thought her wrong; mastectomy is no slight matter.

By late afternoon of the third day one is deep in Western Australia. Now fences and sheep reappear and so do trees — really, low-growing scrub. For once, a kangaroo went in our direction at a distance of a hundred yards or so, racing the train but falling gradually behind. The dining-car steward, whose voice, like that of the head conductor, is piped into all cabins, announced early hours for dinner because the dining-car abandons the train in Kalgoorlie. The conductor then came on to advise that a government inspector would come aboard in Parkeston to "confiscate" all fruit, this being to quarantine against the spread of disease. At dinner our table companion, a gentle, recently retired teacher of music from England stated firmly that the community singing in the lounge car was the worst she had ever heard. My wife, whose judgment I trust in these matters, listened and thought it fine.

At Kalgoorlie, capital of the goldfields, there is a three-quarter-hour pause while the train undergoes yet another change of management — this time to the Western Australia railways. Kalgoorlie expanded from nothing to a temporary peak of 15,000 souls (or the mining-town equivalent) in the six years after gold was discovered in 1892. It then had three daily papers, three breweries, two stock exchanges and a great deal of more sedentary vice. The main street, consisting of one-story, verandahed stores of white-painted wood, is still out of more routine Republic movies. At a classic saloon across from the station, into which we peered, a fair number of local citizens were whooping it up, though in a relatively circumspect way. From Kalgoorlie to Perth on the Indian Ocean it is dark.

On the approach to Perth the conductor called with a cup of tea at six. I had then to contend with an engineering problem no one will have solved even when moon travel is at excursion rates. That is the stuck zipper. When it was released, the suburbs of Perth were at hand. It is a sparkling city, as yet only slightly betrayed by freeways, facing a wide estuary and surrounding a lovely park.

Later that morning the mayor — the Lord Mayor, to be exact — told me he still farmed land rather to the north of the terrain we had traversed. He has a million acres, this being the maximum permitted under legislation enacted some years ago by a leftist government to ensure against excessive holdings. As Malcolm Muggeridge once said, socialism is a trivial thing. However, the mayor's land can sustain a flock of only 40,000 sheep, which means that an animal needs to cover some twenty-five acres to stay alive. This cannot but involve a lot of walking. The train traveler's impression of the distances in Australia must be shared by the average merino ewe.

18

Seven Wonders[1]

I MUST BEGIN this piece with more than the usual number of disclaimers. The most important is that I have never been to Egypt, the Cairo airport apart. My list also excludes natural flora, fauna and waterfalls, the economics of Milton Friedman and the last rites of Elvis Presley. It is confined, in other words, to architectural wonders. And it occurred to me as an economist that my choices might be combined, which can easily be done, thus saving on travel costs. Finally, I am excluding the things that nowadays one sees anyway.

Thus I love to look at the Manhattan skyline at the best hours of the day at the proper levels of pollution. There is, I'm totally persuaded, nothing so wonderful in the whole world as the Ile de la Cité with Notre Dame, the Conciergerie and Sainte Chapelle, and only a step over to the Louvre. But no one can get credit for imagination by committing himself to such available treasures. I travel, as do so many others, partly to arouse the envy of my friends and neighbors who have to stay at home, a feat that is increasingly difficult to accomplish these days.

My first pair of permitted wonders, beginning in the East and coming west, would be the Great Wall and the Forbidden City.

---

[1] A year or so ago *The Sunday Times* of London asked a few pathologically peripapetic individuals to specify the several sights which, were they so empowered, they would now establish as the seven wonders of the world. And to give their reasons. Here are my wonders; a truly astonishing number of letters disputed nearly all of my choices.

The Great Wall, I've been told, is the only man-made structure on earth that is visible from the moon. For the life of me, I can't see why anyone would go to the moon to look at it when, with almost the same difficulty, it can be viewed from China. Everyone has seen pictures of its angled passage up hill and down across the Chinese landscape for its unimaginable 1684 miles. (This figure is from Guinness; there is little agreement on the exact length.) The Wall was not completely successful as to purpose. The Maginot instinct has always been powerful but militarily defective; had more Frenchmen visited the Great Wall, they would have saved themselves much expense and eventual grave disappointment. There is a further detail which should be known to all aspiring tourists. By visible evidence, everyone visiting the Wall has felt an overpowering urge to carve his name on it; Richard Nixon seems to have been the only exception. All who struggle for this perhaps questionable access to immortality should take along some tool for making the requisite inscription. Ample writing space is still available.

I had an excellent mental picture of the Great Wall before visiting it. Of the Forbidden City I had no prior sense. Perhaps some decayed and moth-eaten buildings with roofs that were low in the clichéd Chinese manner. I had assumed also that there would be somewhere close by a stretch of deep sand. This would be where exponents of the kind of discipline that, one gathers, the British National Front now favors, buried criminals up to their necks. A stream of honey then guided ants across this landscape and into the felonious mouth. The ants were expected to finish off the malefactor, though not before he had had time to reflect on his misdeeds, and they performed fully to expectation.

In fact, no such place of punishment was evident, and the Forbidden City itself is a vast congeries of the most fastidious and elegant buildings ever seen. The effect is of great rectangular interiors of wonderful symmetry and perfectly aged and polished wood. So complete is law and order in China that treasures of inestimable value are displayed in these halls with a minimum of protection. Perhaps like that in Britain of the trenches of World War I, there is a lingering social memory of those ants. My further experience of the Forbidden City is of the most courteous, gracious and informed guides I have ever encountered, along with the impression that, whatever the merits or demerits of the Chinese Communists, they did a marvelous job of cleaning up this treasure. It was, I gather, in rather tacky condition when Mao Tse-tung took over. My guide said that several inches of bird dung had to be removed, along with several tons of less nutritious refuse. Other maintenance had been similarly deferred.

The next two wonders, only twenty-odd miles apart, are Fatehpur Sikri and the Taj Mahal. Fatehpur Sikri is, as indeed it has been called, the world's most perfectly preserved ghost town. The houses of the masses were, no doubt, cheap and nasty, and they have disappeared. But the walls, palaces and public buildings are still as when Akbar the Great abandoned them after fifteen years' use in about 1586. They are made of the most elegant building material, a salmon-red sandstone quarried on the ridge nearby. The proportions are perfect and the architectural embellishment superb. When more than a couple of tourists have assembled, a group of exhibitionists hurl themselves off the top of the Gate of Victory into a water tank below. They seem to enjoy the work, and the last time I was there — it was with Mrs. John F. Kennedy, as she then was, when she was visiting India — they excelled themselves in the extravagance of their gestures in the course of their descent. Why Akbar abandoned Fatehpur Sikri is unknown. All of the thoughtful explanations are highly implausible; it is quite likely that the great Moghuls, always restless and of nomadic stock, simply decided to move on.

Agra, close by, was another of the Moghul capitals; it was inhabited for an appreciable period by Shah Jahan, Akbar's grandson. It is an incredible and depressing thought that many people visit Agra and the Taj without being aware of the lesser symmetry but greater grandeur of Fatehpur Sikri close by. I don't suppose there is anything to be added about the Taj Mahal; more than of any other building in the world, it has all been said. Never, especially in architecture, was there a similar tribute by a man of his love for a woman. The literal and physical aspect of that affection is generally affirmed by the fact that Mumtaz Mahal died while giving birth to her fourteenth child. In pictures the Taj Mahal seems rather fine, almost precious; in actual presence it is vast. It also needs to be viewed from early morning until starlight or moonlight, for it is a building of many colors and many moods.

Nothing is more remarkable about the Taj than that it has survived. It is now nearly as perfect in all aspects as when it was completed in its various stages from 1632 to 1650. There was a long time, up to seventy-five years ago, when it was threatened. Some of the semiprecious stones which embellish and enhance the structure were being dug out and removed. Other vandalism was rife. To Lord Curzon, imperialist of imperialists, goes the credit for its preservation, as it does for that of many other monuments in India, including the Ajanta and Ellora Caves, which I would also list as wonders were I in need of more. Curzon is remembered in the history books for his climactic row with Kitchener and the lurking elitism which led to his surprise, perhaps apocryphal, when he visited the Western Front and saw some Tommies bathing; he had not previously realized, he said (or was said to have said), that "the lower classes had such white skin." Perhaps his most enduring monument is, literally, the Taj Mahal.

My next two wonders were, in more peaceful days, within easy range of Beirut. One of them, of course, is Baalbek in the Bal Valley in Lebanon. Perhaps there are more wonderful Roman ruins elsewhere in the world — the Colosseum, for example, or the Pont du Gard. But I doubt that anything exceeds

Baalbek in its combination of grandeur, proportion and beauty. Or better persuades you what truly prodigious people the Romans were. That the eighty-four tall stone columns came from Aswan up the Nile and were rolled or hauled over the mountain from the Mediterranean adds powerfully to the latter point. So does the thought that work continued on the Temple of Jupiter for around two centuries. Not many buildings have been under way in London since the ministry of Lord North or in New York since George Washington. However, both New York and Washington do have cathedrals that are unlikely to be finished for another century or so, money and the building trades being as they are.

Baalbek is magnificent and strong. It is not, however, romantic, and for that one must go north and into Syria to Krak des Chevaliers, which, without any cavil or question (as Julian Huxley averred), is the world's most wonderful castle. It covers the carved-off top of a whole barren mountain, and, in more recent times, a couple of villages have been built out of its walls. The looted stones are hardly missed. However, Krak owes much of its excellent condition to the French, who spent a good deal of time and money putting it back into shape during the years of their Syrian mandate.

Krak des Chevaliers was garrisoned by the Knights of the Order of the Hospital of St. John. These were the armed monks who began as the protectors and healers of pilgrims going through the Islamic wasteland to the Holy City and then became the sword arm of the Crusades. Then, after being driven from the Holy Land and expelled from Cyprus, they combined pious works with diligent piracy from the island of Rhodes. Eventually they went on to Malta. Through the great angled gate of Krak des Chevaliers, a whole company of the mounted monks could gallop in or out at full speed. A majestic curtain wall surrounds the castle proper; in the keep is a lovely chapel and a fine vaulted chamber which housed the local head of the order. We visited Krak one cold rainy spring day in 1955 and were at the time the only tourists within miles and, I would judge, days. We were shown around by an aged Arab who made up in hospitality what he lacked in personal hygiene. At the end of our tour he took us to his quarters and a warm fire. There he told us that he had learned his English in Montreal where he had worked for the Canadian Pacific Railway. I was at that time a consultant for the CPR, which had recently sought the guidance of a group of Harvard economists in, among other things, using its new Univac computer for a study of its more egregious costs. I was able to bring him up to date on life at the railroad. He told us how he yearned to see Windsor Station in Montreal once more. There is no accounting for tastes.

I have many candidates for my final and odd-numbered wonder. The dead cities in Ceylon are interesting but do not succeed in competition with Fatehpur Sikri. There is the Black Pagoda at Konarak on the Bay of Bengal just north of Puri in India. It is a great stone chariot mounted on beautifully executed wheels and drawn by a team of gorgeous horses. It is covered with exquisite carvings showing the kind of sexual recourse that my generation once associated exclusively with the Place Pigalle. I first visited Konarak in 1956 with the late John Strachey. John looked at one couple who had sustained a highly calisthenic embrace since the mid-thirteenth century (maybe longer), drew a deep breath and said, "Jolly good! *Jolly good!*" I hadn't previously realized that British M.P.s and former cabinet ministers could express such depth of feeling with such eloquence. Had Cambodia been left in the obscurity and peace it deserves and for which its people unquestionably yearn, I might have managed to get to Angkor Wat. That never happened. Nor did I ever get to Petra. Once when I was in India, the new Jordanian ambassador, a most agreeable and jovial man, asked me to stop over in Amman and promised he would take me there. Unfortunately a day or two later he fell afoul of the Indian currency exchange regulations. As he was entering the country, he had included, in a thoughtless way, a trunk or mattress full of gold with his household possessions, this being marketable at a premium in India. Whatever it was broke open in Palam Airport, and the gold made an impressive sight on the floor. In consequence, he was held *persona non grata,* and I thought it a little uncouth to ask him to keep his promise.

I am left, since I wish to be suitably esoteric, with Machu Picchu. The actual ruins themselves are impressive. So is the thought of the primitive stone tools by which they were accomplished. So, above all, is the site. The city stands on top of a high Andean mountain which is shaped like a greatly attenuated beehive. One looks across the endless green of the jungle and down to the fast waters of the Urubamba River far below, and one reflects that this city, with its temples, walls and houses, remained undiscovered until 1911. In that year Hiram Bingham, believing more or less in the legend of the lost city of the Incas, came upon it and had all doubt dispelled. Bingham, I should say, was a Yale professor of Latin American history and later a United States senator from Connecticut. He shares with Joseph R. McCarthy the distinction of being one of the few senators in our history who was ever formally censured by that body. Bingham's crime, which in modern times would be considered a commendable exercise in legislative diligence, consisted in bringing a business lobbyist from his state into the Senate Chamber to advise on an impending tariff bill. His son, Jonathan Bingham, is a former ambassador and now a senior and highly distinguished member of the House of Representatives from New York. This has nothing to do with Machu Picchu, but no one going to South America should miss it. And indeed not many will. South America is singularly barren of architectural wonders. If these are one's interests, one should go there after the tenth visit to Europe, Asia or the Middle East, the fourth (if one is a European) to the United States and just before going to Australia.

19

Circumnavigation 1978

September 10 — Sunday

WE ARE TRAVELING around the world. The direction reverses
Sir Francis Drake, but the purpose remains the same — exploration, adventure and pillage, the last directed not at Spanish
galleons but at lecture audiences in both Europe and Asia.
Mark Twain had the same view of lecturing I do: "A man can
start out alone and rob the public, but it's dreary work and a
cold blooded thing to do." I have also in mind observing some
of the current generation of statesmen in action, revisiting old
scenes in India and seeing friends in Thailand. Also doing a
great many things in Japan. There is no alternative to doing a
great many things when in Japan.

My publisher in Tokyo, IBM and Management Centre Europe are paying our way, with some overlapping revenue to the
IRS. In consequence, we are traveling first class. At the TWA
counter at Kennedy, where the turmoil was barely supportable,
the line at the first class counter was the longest of all. Maybe
the income pyramid is at last inverted. Wider at the top; better
to be poor. This I doubt.

Once we were on the plane, an engine required repair, and
the TWA 747 scheduled for 9 P.M. was two hours late out of
New York. The pilot has now twice told us the flying time will

be around seven hours to Milan. This is very troubling, for we are supposed to be going to Rome.

September 11 — Monday

Leonardo da Vinci Airport was as congested as Kennedy but at a lower level of decorum. In my liberal youth I thought how good it would be if everyone could have a vacation in Europe. Now all do. Two lone immigration officials examined passports and checked names against a list of known terrorists. Surely any sensible known terrorist would carry false identification.

The Hotel Hassler on top of the Spanish Steps remains, to a limited extent, a haven of rest. This could be because our room, of modest size, costs $110 a day — to my publisher, fortunately. I detect another interesting economic tendency. The affluent in these days of mass travel seek seclusion. This the best hotels ensure by setting exorbitant rates. The more they raise their rates, the more seclusion they claim to offer and hence, no doubt, the more customers they have.

At dinner there was mention of the peace that has descended on Rome. The Red Brigades are thought to have gone on vacation. Gore Vidal and I exchanged thoughts on publishing in the Soviet Union. He is currently the most popular American fiction writer there; my books are not seriously competitive. We both find the Soviet editors meticulous. They make few changes and are careful to ask about all alterations. Luigi Barzini and I discussed work. We agreed that what we do, namely write, should not be confused with physical toil; that all societies invest great effort and emotion in propagating the myth that good solid physical labor is ennobling, something that every normal upright person seeks, glories in and enjoys. There is no ethic like the work ethic. Then all who can, make their escape to physically less taxing occupations and carefully pretend that these are work too. Vidal said he writes two hours a day. I write about four.

September 12 — Tuesday

Today I had a long discussion on television about the Italian economy with leaders or near-leaders of the Christian Democratic, Socialist and Communist Parties. Like the British economy, that of Italy, which seemed hopeless a few months ago, now seems quite promising again. The inflation rate has fallen sharply. Incredibly, and at least momentarily, the balance of payments is in surplus — tourist revenues, overseas remittances and small business exports are all high. Some capital that took flight when it seemed that the Communists might come to power is thought to be returning.

I was asked repeatedly if I would approve of the Communists being part of the government. I replied that my approval is not necessary. However, I would approve. Modern industrial development disperses power to many claimant groups. Where once capitalists confronted workers, there are now managers, intellectuals, trade unions, organized farmers, civil servants and many others, all demanding a voice in public affairs. So the monopoly of power by capitalists or workers as envisaged by Marx is no longer possible. A legislature becomes the only way of arbitrating as between the different claimants when none can be denied. I was pressed as to how general was this view in the United States. I said that no one should assume that the basic Cold War mind is the best we have; it is merely the most fixed.

In the early evening I visited Gianni Agnelli at his flat; various members of the Italian Establishment dropped in. Agnelli is the most intelligent businessman I know. Exceptionally among executives he does not believe that he runs his business by divine right, and he refuses to express any indignation as the scope for entrepreneurial decision narrows around him. Rather, he always seems a little surprised that Fiat has survived so successfully for so long. The talk was of economics, and someone suggested that Italy was turning the corner because it was being discovered that hard decisions, especially on the budget, could be popular. Greatly needed is better management of the publicly owned industries, of which those in Italy are extensive.

On politics, I was asked if Edward Kennedy saw himself as an alternative to Carter and if he wasn't deterred by the fate of his brothers. Agnelli intervened to suggest that danger could be interesting. I noted that the Kennedys didn't tell their friends their secrets, for they had learned that their friends didn't keep them.

September 13 — Wednesday

I had to cancel a meeting with Enrico Berlinguer, the head of the Italian Communist Party, in order to be sure of getting my plane to Nice to speak to a gathering of IBM executives. The inevitable triumph of capitalism.

Anyone leaving Rome should allow half a day for doing so. Long lines at the ticket counters, longer ones at passport control, a terrible jam before the baggage-viewing machines. Though I cleared the ticket counter with forty-five minutes to spare, I barely made the plane.

The Nice airport was much better, and the Riviera was covered with a soft, hazy autumn sunlight mixed with smog. I had dinner with an eclectic gathering of the heads of IBM in various countries, their presence here being a reward for exceptional achievement in computerization. During the meal my dinner companion, a woman of agreeable aspect, told me that she had once worked for IBM but had quit the business when she married Ralph A. Pfeiffer, Jr., our host and the head of one of the international divisions, who was present in an overpowering pair of pea-green pants. Since retiring to her family, she had interested herself in theater and communications. I listened in a condescending way and am glad I did, for there were toasts later in the evening celebrating the day's announcement in the *Wall Street Journal* that she had just been named Chairman of NBC.

September 14 — Thursday

A lovely day on the Riviera — warm sun, locally still some soft smog. The IBM seniors listened with attention to my lecture and avoided the questions that normally turn such occasions into disaster, "Professor, wouldn't any restraint on prices and incomes be the end of free enterprise as we know it?"

After lunch I went with a friend to see a huge Giacometti exhibition at the Maeght Foundation at Saint-Paul de Vence. The gnarled, enormously attenuated sculptures are so expressive that it is possible to survive their profusion. This is not so of the drawings and paintings. Some day some heroic museum will have an exhibition at which only the ten best things of an artist are shown. More than fifty involve not a diminishing but a negative return.

The Maeght is on a hillside looking distantly toward Nice and the Mediterranean. It's an intricate combination of garden space for sculpture and small glass-sided galleries, these on several levels — the work of my Cambridge neighbor José Luis Sert. It is lovely, and so are the Miró ceramics, which it features. A museum official identified me and thought it "nice that someone came from Cambridge to see Sert." I'm staying overnight in Auribeau in a tiny house of great delicacy of taste and line, Auribeau being a small village of medieval provenance on top of a high hill between Grasse and Cannes. Cars must be left outside; within, steps replace streets or even footpaths. Sublime.

September 15 — Friday

In the morning a walk of several miles along the Siagne in perfect weather. Then by car to Cannes and along the Mediterranean shore road to Nice. The waterfront is now indistinguishable from Long Beach, California, but rather more hideous. Women of good appearance approach the sea in a general way without bathing-suit tops, which is sensible, cool, attractive and, I would judge, economical. Air Inter flies beautiful big planes

of an undisclosed type from Nice to Paris. The color scheme
is pleasant; there is plenty of leg room; an excellently simple
meal is served with efficiency and dispatch. The French airports
are busy, but the crowds move with well-synchronized discipline.
I'm reminded once again of Nancy Mitford's comment that the
French manage well everything the British are supposed to
manage well.

September 16 — Saturday

I have two books being published almost simultaneously in
Paris so I spent nearly all the day in literary self-praise. The
method is to combine unconvincing modesty with an underlying commitment to extreme worth. I also did an interview
with Walt Whitman Rostow with whom during the Vietnam
years I had deep disagreements. But I judge him, unlike others,
to have put much of his past behind him. He has not joined the
Committee on the Present Danger, in which the erstwhile warriors preserve their hopes for the new war that will retrieve
their reputations for foresight and alarm. Anyhow, there are
better things than stroking old wounds. We spoke of the Carter
administration, the need to cut oil imports and get incomes and
prices under control and the tendency of administration economists to remain balanced between the fear of action and the
fear of the consequences of inaction.

In the afternoon I had a long interview with *Paris Match*.
When my answer pleased my inquisitor, he wrote it down.
When it didn't, he ignored my response. A man of character.

In Paris I avoid the memorable restaurants. That's because
while talking, I rarely notice what I'm eating so I spare the
expense. But Nicole Salinger, whose long interview with me
had made one of the books I am here to celebrate, persuaded
me to go to Maxim's. A wandering violinist came by and asked
me for my favorite song. Alan Jay Lerner was at the next table
so I suggested that the musician play extensively from *My Fair
Lady*.

September 17 — Sunday

Sunday and more self-praise — this for French radio. My interlocutor asked me not to mention economics. "That," he said, "causes people to turn off their sets." Since the purpose was to interview me — and Nicole — on a book on economics, the restriction seemed a trifle confining.

At noon I had lunch with Maria Teresa de Borbón Parma, a woman of striking, dark-haired, dark-eyed beauty and a direct, if slightly distant, descendant of Louis XIV. Her brother is (or was) the Carlist claimant to the Spanish throne; like her brother and sister, she is a convinced democrat and socialist. After one hundred and fifty years, more or less, of intense hostility, the Carlists are now on speaking terms with the legitimate or, anyhow, more successful line. She is rather admiring of Juan Carlos's skill in guiding her country back to democracy, although she worries about lurking intransigence in the army and on the extreme left and right. With her brother and sister, Maria Teresa is standing for parliament in the next election, always assuming that she (and they) get their citizenship back first.

This evening I walked with a friend along the Seine. At one of the booksellers a man was reading Henry Miller. His companion, quite lovely, was reading Raymond Vernon (of Harvard) on multinational corporations.

September 18 — Monday

This morning's duty was a three-hour meeting with Japanese businessmen who are touring Europe as the "Study Group on World Economic Forecast." About thirty were involved; with translation and questions, it took a full three hours. With travelers of any other nation I would assume that the education was a disguise for tax deduction. Not my students. All took extensive notes to supplement personal tape recordings and an overall transcript. Questions were intensely practical and frequently

flattered my competence. There was much discussion of the
point on which Barzini and I had touched — is it inevitable
that as industrial countries mature, people will work less hard
and increasingly reject tedious, repetitive work? I so held and
noted that the tendency had been partly concealed in the United
States by drawing on new drafts of labor eager to escape the
worse life and greater toil of Appalachia, the Old South and the
Puerto Rican cane fields. And in Europe it has been disguised
by the heavy use of foreign labor. Japan, a young industrial
country, has not faced this problem. Some day it will.

Lunch was with Claude Gallimard, my exceptionally talented
and attractive French publisher and his equally agreeable staff.

In late afternoon my taxi tackled the traffic to Charles de
Gaulle. Quite unexpectedly we broke out of what seemed to be
the ultimate tangle and arrived.

September 19 — Tuesday

Copenhagen is cool and sparkling clean. My housing in the
local Sheraton is low-level plastic. Since I was last here two or
three years ago, pornography seems greatly to have declined.
I've long been persuaded that sex is commercially successful
only when illegal or immoral; otherwise it lacks cachet and
attracts too much everyday talent.

I spent the day lecturing businessmen on behalf of Management Centre Europe, an organization devoted to the greater
enlightenment of entrepreneurs. I was the keynote speaker, although the other speakers did not respond audibly to the note
I struck, if any. The most urgent question by the business audience: "Should a well-run European firm have a foothold, i.e.,
should it own a business, in the United States?" The answer
from all my European fellow sages was an unequivocal "Yes."
There was a definite impression that while all European countries are in varying measure socialist, the United States remains
devoutly capitalist. I pointed out that social legislation in Europe is enacted to advance socialism, while the same measures

in the United States are put into effect to safeguard the free enterprise system.

September 20 — Wednesday

I had lunch today with Warren Manshel, the newly arrived American ambassador. A financial man of sorts, he was a stalwart opponent of the Vietnam war, and it was then we became friends. It's hard to see how anyone with such a past qualifies for a high diplomatic position. We usually prefer men who are willing to show, even at the expense of being wrong, that they are capable of taking a hard line with people of peace-loving tendencies. Several newspapermen joined the lunch, which, rejecting all precedent, was not at the Embassy Residence but at an excellent downtown restaurant. They asked, as usual, about Jimmy Carter and why his economists were doing so badly. I explained that under our system of upward failure the economists who do worst, get the most publicity and thus personally do best. Alan Greenspan and William Simon first inspired and then guided Gerald Ford to the economic policies that lost him the election. Both went on from this disaster to careers of great distinction as a result.

In the evening to Brussels.

September 21 — Thursday

Another long day of lectures to business executives, again for Management Centre Europe. Much of the interest was in the proposed monetary union for the Common Market. French francs, Deutsche marks, lire, Belgian francs, will remain and all will have a stable relationship with each other. This association is called "the snake," reflecting the general incapacity of financial men for metaphor. Supported by an articulate Swiss banker, I held that it had not the slightest chance of surviving with any binding relation between the parts as long as different countries have different wage, price, fiscal and monetary policies and thus different rates of inflation. An alignment of internal policies must come first; until this is done, currencies will fluctuate — or, at a minimum, the snake will periodically come apart. The audience mostly disagreed. Businessmen, even more than politicians or the simpler kind of economist, want to believe that there is some as yet undiscovered magic in the management of money.

September 22 — Friday

A very quiet day in Brussels. Its Sheraton is less reprehensible than that in Copenhagen. Lunch was with Léon Lambert, who lives with lovely paintings (and more Giacomettis) on top of his own bank. Attending were senior Belgian politicians and businessmen, the Establishment. An aide to the King rebuked me gently for neglecting to call on His Majesty. I promised improved manners henceforth. The talk was of the future. I have reached the age where I comment on this with confidence, for I won't be around to hear about it if I am wrong. What will be the major anxieties ten years hence? Some kind of rapport will have been established on incomes and prices and the regulation of public and private expenditure so that inflation will not be the primary problem. In Europe there will be much concern over regularizing the role of the mass of foreign workers. No one should imagine that they can be kept forever as a special subproletariat. All the older countries will be reconciled to the departure of the simple, tedious industries to the more competent countries of the Third World. Steel, heavy chemicals, tires, ordinary textiles, shipbuilding, will be gone. (Steel is already in deep trouble in the United States, France, Britain and the old districts of Belgium.) The older industrial countries will still have computers, aircraft, missiles and other advanced weapons of mass destruction. And if they survive their excellence in the latter, they will have the industries which require good or original art and design.

September 23 — Saturday

I left early this morning and was warned before departure of various airline strikes. All went normally to Frankfurt, and

Pan Am 2 to Tehran and Delhi was late only in the normal way. The Pan Am people were greatly surprised to see me boarding, all reservations having been accidentally erased. I became intensely disagreeable, which, when righteously inspired, is true joy. I remembered and reminded the local official that they only flew into various Indian airports because I had got them landing rights. Later it occurred to me it was TWA I had helped.

September 24 — Sunday

Anarchy, if sufficiently unrelieved, can become a mild form of business genius. Pan Am started the movie so late it was only half finished when we reached Tehran. Thus all disembarking there will have to take the same plane back in order to see the ending. They then ran it right through the forty-five-minute stopover, thereby keeping the through passengers from contemplating the revolution outside.

This is my first significant return to New Delhi since I served here fifteen years ago. (I came once to represent L.B.J. at a funeral, and I passed through once when raising money for the Bangladesh refugees.) Nothing is more inconvenient for an ambassador than to have a predecessor on the premises, so we had arranged to billet ourselves at the Indian International Centre. But the Goheens, being absent, had us moved to the Roosevelt House, the majestic Edward Durell Stone creation of my years and the official Embassy Residence. I named it for F.D.R. and, to be sure all knew it was for him, got Averell Harriman to donate a Jo Davidson bust for the entrance. The Republicans, when they came, dug up a bust of T.R., I would judge from a Long Island antique shop. I never heard that T.R. had much to do with Indian independence.

Tonight we dined with the head of the Foreign (External Affairs) Office, an old friend. We have still to go, with Richard Roth (the movie producer), to a celebration on behalf of lions, tigers and other wildlife. In a country where so many people have so little to live on, one must be cautious about expressing too much concern for animals. I remember a woman in Calcutta straight out of Evelyn Waugh who told me that it couldn't be good for the cows to be loose on the streets with so many people.

September 25 — Monday

This morning I called in at the Embassy Chancery, still perhaps the most beautiful building ever accomplished by the government of the United States — an elegant rectangular shell surrounding a beautiful water garden. I was reminded of taking Lyndon Johnson into it in the spring of 1961. He said with indignation, "What did this cost?" But I knew L.B.J. of old. He was a master of intimidation, and the remedy was homeopathic. I said sternly that it had, indeed, cost a great deal and was worth every cent of it. He agreed.

At noon I met with numerous Indian editors and publishers, and we discussed the relative decline of the superpowers since my years in Delhi. At that time, the two giants towered over all others; Indian foreign policy was that of the nervous embrace. It wanted the considerable rewards that came from attempted seduction by one or the other without the seduction.

Now much has changed. China challenges Russia in the Communist world. The vision of a unified Communist conspiracy has been abandoned, it is said, even by James Angleton.[1] On the other side, the industrial eminence of the United States is powerfully challenged by Germany and Japan. As the perfectibility of socialism has given way to worry over the problems of management and especially of agriculture, so the wonders of free enterprise in the age of Keynes have given way to worry over inflation and unemployment. Only Walter Wriston has yet to hear that capitalism has problems. Meanwhile Vietnam showed us the limits of our power in countries geographically distant and culturally different from our own. In China, Algeria, Ghana, Egypt and Somalia the Russians have

---

[1] See page 336.

been taught, and one hopes have learned, the same lesson. I emerged without damaging challenge. Some of my audience, I sensed, still prefer a world in which foreign policy involves only a choice between two superpowers.

I had a long talk with Aisha Jaipur, in my time the elegant Maharani of Jaipur, who in recent years has had a major stay in the stony lonesome. Government mine detectors, as I recall, uncovered a major deposit of metal in the environs of one of the Jaipur princely palaces. It turned out, disastrously, to be undeclared gold. Possibly it had been buried there in past times and forgotten. Aisha didn't care much for the experience. She also regretted some errors in her autobiography (written with Santha Rama Rau), which was published while she was in the clink. The proofs, she said, had to be smuggled in and out and read surreptitiously. That was bad for accuracy.

September 26 — Tuesday

The Indian Airlines plane to Kashmir operates with a 90 percent load factor in summertime, has a forty-five minute turn-around time in Srinagar, and, I'm told by the local manager, makes a great deal of money. Kashmir is one of the few disputed territories in the world that is worth the trouble and expense. The others are places that civilization reached last, and for good reason.

Here we are staying with L. K. Jha, once a fellow economics student in England; later, as Secretary of the Ministry of Finance in New Delhi, my constant companion when we were extensively the underwriters of the Indian balance of payments; then an exceedingly popular ambassador in Washington. Now he is Governor of the state of Jammu and Kashmir. The Raj Bhavan or Government House is on a hill overlooking Dal Lake, the famous floating islands, the near and distant mountains and the Srinagar television tower. A pleasant, rambling wooden structure, it was built by the last Maharajah before independence, when he was advised by competent astrologers, who must have been closet democrats, that he could only have a son if he married a woman of low birth. This he did, but since dignity, as distinct from democracy, did not allow him to go to a low house for the wedding, one had to be built in decent grandeur especially for the bride and the ceremony. The original palace is a few hundred yards away and is now an excellent hotel. At this season in Kashmir flowers are everywhere.

The state of Jammu and Kashmir lies on the two sides of Banihal Pass. Kashmir is high in the mountains; Jammu is on the hot plains. Snow comes to Kashmir in the winter; Jammu is pleasantly warm. Accordingly, as autumn recedes, the state government, down to the lowest clerk and the most neglected file, moves to Jammu. This takes about a week as compared with the weekend required in British times to take the entire government of India from Delhi up to Simla.

September 27 — Wednesday.

A day in the Vale of Kashmir and lovely beyond easy description. Rice harvest was in progress. The grain is cut by hand; women and teenagers carried away prodigious loads of straw on their heads. A team of two cows — not oxen — was plowing the stubble with a wooden plow unchanged, I would judge, in the last five hundred years. Unchanged also are the Shalimar Gardens, going back to the Mughals, and a major riot of red, yellow and purple. We had lunch at a game sanctuary, once the private hunting preserve of the Maharajah. There is some tension among the animals in the enclosure. The snow leopard recently bit the leg of a rare and valuable species of deer. The leopard died, and the deer survived. On Dal Lake, as we returned, the *shikaras* (gondola-like boats) were out, and people were pulling up the reeds and other water debris that make the floating islands that grow vegetables. On the way to Srinagar we paused to visit Claremont Houseboats, G. M. Butt proprietor, who has numbered among his clients everyone of official importance who has come to Kashmir in the last forty

years. Pictures of all remain. Mine is next to Nelson Rockefeller's; Nelson's, however, is in color. Then we took a long
shikara ride down the Jhelum, the main river of the valley
passing through the city of Srinagar, which rises densely on
either side for two or three miles. Then we drove back through
the city. In most places handicrafts are something that people
who combine nostalgia, artistic sensitivity and a certain excess
of funds seek, unsuccessfully, to encourage. In Srinagar every
kind of hand manufacture — carpets, carving, papier-mâché,
copper jewelry, furniture, footstools and more, ad infinitum —
has an explosive life of its own. Thousands of shops proclaim
the resulting wares. One advertises "thrilling" shawls.

A large sign in the center of Srinagar would have usefully
warned Richard Nixon and Spiro Agnew:

INCOME TAX EVASION IS A DISGRACE
IT IS ALSO ILLEGAL

September 28 — Thursday

The day was spent in bed with a sinus attack. Dinner tonight
was with Sheikh Abdullah, the Chief Minister, and a small
party including the Vice-Chancellor of the University. (There
are two state universities, one here and one at Jammu.) Sheikh
Abdullah, the Lion of Kashmir, is a most impressive figure,
very tall and solid in proportion. He was for long the advocate
of independence for Kashmir. In consequence, he lived for
many years under one form of restraint or another. Now he has
become the voice of moderation against the irreconcilables who
still want to be part of Pakistan.

We talked mostly of economics; the growing population of
Kashmir, up by 50 percent since independence; the booming
tourist trade and the need to build more hotels; the need to
keep houseboats from disfiguring Dal Lake; the boom in handicrafts, in which Srinagar leads the world in extent and variety;
the need for more money to finance further development.
Withal, the people of the valley, or most of them, remain dis-

mally poor and must also, unlike the people of the plains, contend with a cold winter. At the lower levels of poverty they don't heat their houses; instead, as of old, they carry a small firebox of live coals around under their clothes. It is an acute carcinogen. Skiing is a growing industry.

Sheikh Abdullah told of having to go to the university to tell the students to take an examination to which they were objecting. Their protest was on the usual political grounds — the test was too hard and thus discriminated adversely against those who were not by nature bright. Most of the universities in northwestern India are in poor shape; some ran as few as fifty days last year and were closed the rest of the time by one form of agitation or another. The agitation is not without purpose; a student who combines ambition with a deep allergic reaction to books becomes an agitator. This then marks him out as a political leader and a potential legislative candidate when he graduates. It is a kind of natural selection of the worst. The resources for running India's huge university system come out of a very poor community. I was always angered that the students, a highly privileged group studying at the public expense, were allowed to make such bad use of such scarce resources.

Kashmir has two disputed borders. One is with Pakistan, where, in the main and as regards the best land — the valley — the Indians are in possession. The other is in Ladakh, where the Chinese hold the actively disputed territory. Time is gradually resolving both disputes in favor of the tenants-in-being.

September 29 — Friday

On this golden morning we went shopping in Srinagar, a joyous exercise, for at this time of year in any one of the thousand shops you are the only customer, perhaps the only one of the day, and welcomed accordingly. I called in on the famous emporium of Suffering Moses, and the owner told me he wasn't selling anything; he was taking stock to see what, if anything, had recently been stolen. A tailor, Mr. Kahn, from whom I bought a jacket sixteen years ago, recognized me on the street, asked how the garment was wearing and as to my need for a replacement. At noon we caught the plane back to Delhi.

There we went to make a courtesy call on the President, Sanjiva Reddy, whom I knew years ago as Chief Minister of Andhra Pradesh. There is a style about such ceremonies in India that is sadly lacking in Washington. For one thing, the Rashtrapati Bhavan, in its red-sandstone, Mughal-garden magnificence, dwarfs the White House approximately as the Empire State Building dwarfs Altman's. For another, the aides who greet, salute and guide you into the labyrinth are to the White House policemen as the Rockettes to Radcliffe freshmen. No wonder Richard Nixon wanted something more Mikado. Our conversation was extensively reminiscent.

Then I went to the Indian International Centre, a handsome collection of buildings adjacent to the Ford Foundation and the World Bank in the Lodi Gardens. There I gave a lecture, and the hall was crowded, including all standing room. The tribute is to the affection with which I am regarded in India and which I find so pleasant that I greatly wish I had cultivated it in the United States. The lecture was on the economic problems of the developed countries. Most economists visiting India feel obliged to speak on the problems of the poor countries and what should be done about them. This is deeply insulting. Also the advice is usually either wrong or politically or administratively inapplicable, and, in any case, it is disregarded.

From the lecture we went to dinner with a small cross section of the New Delhi Establishment — old friends, lovely wives. The evening's discussion turned to the stability of Indian democracy. There was a variously expressed feeling that Indians, influential and less so, have a strong sense of participatory power in their government and are determined not to lose it. Thus the surprise at the defeat of Mrs. Gandhi is readily explained: some millions took the thoughtful precaution of expressing themselves in her favor before going to the polls to vote against. I long have had the further feeling that, in a country not exactly replete with recreational opportunity, politics is a source of much enjoyment. Speeches, promises, acrimony, scandal, victory, defeat, change. This no one or not many wanted to lose.

All contemplating a trip to India should think of late September. Warm days, cool nights, all green after the monsoon. None of Kipling's pile of sand under a burning glass.

September 30 — Saturday

This morning we went to visit Prime Minister Morarji Desai. We were shown to corner seats in the large living room of his residence so that pictures could be taken with a maximum of convenience. He came in looking a couple of decades younger than his eighty-two years, certainly no older than when he was Finance Minister under Nehru and I used to see him every week.

Our talk ran first to his recent visit to Moscow and the desire of the Soviets for a SALT II agreement. He then expressed his warm approval of Cyrus Vance and Jimmy Carter and asked me how I now saw Indian relations with China, in which I was once (at the time of the 1962–63 war) considerably involved. I urged the need for letting the boundary dispute in Ladakh (beyond the Himalayas) lie fallow. The real estate involved is uniquely barren; what could not be settled sixteen years ago could be more easily settled now, and in another ten years passions would further cool. He demurred, adverting to political pressure for getting the Chinese off Indian soil, however distant and inexpensive. We agreed in our distaste for any inclination to make capital from the tensions between China and Russia. He said that he had pressed the point on Jimmy Carter, who also agreed. He asked me if there was some tendency to the contrary in Washington. I pointed out that we had always to contend with those who, having read the books, saw themselves in the mantle of Machiavelli, Talleyrand or J. Foster Dulles.

He said that he had little trouble in identifying the person or persons so motivated. Altogether it was a pleasant time.

We had lunch with a large group of journalists and civil servants, and the discussion was on North-South relations, as all intercourse between rich and poor countries is now described. This is an area of great surprise. The rich countries strongly encouraged the industrialization of the poor countries as an unexceptionable good — who could be against economic development? Now the question of buying the resulting textiles, steel, steel products and other industrial goods has come up. That, naturally, no one expected; these things were all supposed to be consumed at home. It is the weakest industries in the advanced countries and those with the highest labor content and cost that feel the competition. So protectionist sentiment is reinforced by compassion. It will be an interesting cause of tension in times to come.

Then came an interview with the *Hindustan Times* by a sensationally good-looking woman, who began by asking me my I.Q. I told her truthfully that I had never had it measured. She left me wondering. She asked how we raised our children and why we had remained married so long. I enlarged on the need to devote as little time as possible to one's offspring lest they acquire one's bad habits; to one's virtuous tendencies, if any, they are naturally immune. I held also that marriage being a perilous and improbable association, it is safe only if the principals don't see too much of each other. She wasn't altogether persuaded but copied it down.

Then I did a long interview on All-India Radio — interesting, fast-paced questions which I much enjoyed — and this was followed by a dinner by the Government of India. The latter, of much style, was at Hyderabad House, a place for official entertainment which is on the same scale as the Metropolitan Museum and is the onetime Delhi headquarters of the Nizam of Hyderabad. Journalists, senior civil servants and a few politicians were there. The speeches and conversation were mostly

of a lighthearted sort. I was queried on the tendency of the
Asian subcontinent to what I once called the North American
solution. The doctrine in question holds that North Americans
are no less belligerent than normal and could be more so, but
peace prevails because the British and Spanish left behind one
big country with a rim of smaller ones. The big country is
morally restrained from attacking the small ones; and no American politician can make much capital by assailing Canadians,
Mexicans or Guatemalans. Cuba proves the point; Cubans became bad only when they came to seem dangerous. The Canadians and Mexicans in turn have become accomplished in living
beside a big country and getting out of their situation as much
independence as they can.

The same solution should have emerged on the Indian subcontinent — India and a fringe of much smaller states. It was
delayed by the perverse genius of the American global strategists led by Dulles who sought to build up Pakistan as a military ally. That made attacks on Pakistan a profitable occupation
for Indian politicians and led on to three wars. Now, with the
independence of Bangladesh, the shrinkage of the relative position of Pakistan and the decline of the global warriors in Washington, the subcontinent has settled into the North American
equilibrium. The theme served for an hour or so without encountering serious objection, save for some words of caution as
to what the Pakistanis might one day yet do.

October 1 — Sunday

Nothing of importance transpired at lunch or later at a picnic
of incredible charm on the edge of Delhi. An estimated eight
cities, some say seven, have occupied the approximate site of the
present capital. Where we went is a once densely inhabited,
vast archeological area called Mehrauli near Qutb Minar. Our
picnic was on the top of the house and tomb of an early six-
teenth-century poet named Jamali and his brother Kamali.
Ruins, all unexplored, stretch endlessly around. We watched
the sunset from the flat balustraded roof.

In the evening we had dinner with Rajeshwar Dayal, who was for many years a senior figure in the Indian foreign service and who, in the days immediately after the independence of the Congo, was sent by the UN to rescue it from civil war and the even more dangerous attentions of our cold warriors. I was Dayal's conduit to President Kennedy, who was highly sympathetic. I talked at length with S. Mulgaokar, the editor of the *Indian Express*. I held that the Indian newspapers, although contentious and lively, devote far too much attention to political speeches and political news in general. If a politician speaks but says nothing, only this fact should be mentioned. Mulgaokar partly agreed but insisted that Indians have an inordinate appetite for political news. Speeches are also cheap to report, and politicians have something to do with the supply of newsprint and advertising. (This morning's *Hindustan Times*, generally thought the best in Delhi, had a three-column headline announcing that GANDHI COMBINED IDEALISM AND PRAGMATISM. Prime Minister Desai was speaking on the eve of Gandhi's birthday.)

We went to bed at 11:30 P.M. and got up at 4 A.M. to catch a Lufthansa flight to Bangkok.

October 2 — Monday

This is my first visit to Bangkok in seventeen years — almost to the month. President Kennedy sent me out to Saigon in 1961 to make a report after I had expressed myself against the Maxwell Taylor/Walt Rostow paper urging greater military intervention in Vietnam. Troops were to go in disguised as flood control workers. I stopped in Bangkok afterward to send on my views. That was because the Saigon embassy and ambassador were pressing strongly for troops, and it seemed a little crude to file from there saying they were full of it.

Bangkok, with a population of over four million, is now enormously larger, and the resemblance is not to the old city of the canals but, generally speaking, to Atlanta. As always, the people look infinitely more prosperous than those in India, as indeed they are. A former Harvard student of much charm met us at the airport after arranging for the elision of all entrance formality. We are housed in a fine suite on top of the Hotel President, which her husband owns.

The most talented man in Southeast Asia is Kukrit Pramoj. With some royal connections, he is also a newspaper publisher, journalist, a student of Thai dancing, music, art, antiquities and architecture, and a former prime minister. He lives in an assembly of ancient Thai houses — long, with steeply curving roofs like those of the Forbidden City. Within, there is a glistening vista of polished teak. All is a museum of Thai art and culture. We went there for dinner. Beside the dining room is a small theater, and back of that for the evening was a small orchestra. With dinner we had Thai music and a succession of classical Thai dances by students and friends of Kukrit. It was a major evening. There was only passing mention of Thai politics. Kukrit did say of one cabinet officer, a man of some education, that he threw cultured pearls before real swine.

October 3 — Tuesday

This morning I gave my inevitable speech, this to a combined assembly of all the various professional economic organizations of the city. The questions were sensible and thus unnotable. Then we went to a superb small museum in the house of Princess Chumbhot, a woman of charm and power in the land. She greeted us, as did a huge pelican which accompanies all visitors around the grounds. We saw a smallish temple of much elegance, replete with shrines and artifacts and some exceptionally durable pottery dating to 4000 B.C., which is thought by many to establish Siam's priority in this branch of art and manufacture. The point seems well taken. Also some excellent old boats of strictly ceremonial aspect.

October 4 — Wednesday

This morning we toured the adjacent city and countryside by boat. We went down the Chao Phraya River heavily in flood and then circled through canals in what was once the great orchard area of Siam, to return to the river considerably upstream from our starting point. All life — houses, shops, shopping centers of a modest aquatic sort, temples and schools — is related to the water, and everywhere children were swimming in it without apparent damage. Some strength must be needed to shove through. Mark Twain said that old Mississippi river hands, when taking a drink, stirred up the sediment in the river water in order to be sure of getting the nourishment. Here it would not be necessary.

At lunch we did get around to politics. A constitution is being drafted; it is believed to promise that half the parliamentarians will be elected, half appointed by the government. Some younger people present saw in it a chance for a civilian front to win power from the army. Kukrit was not attracted by the idea. The present government is corrupt; the new arrangement would merely share the booty.

In the Buddhist lands, it should be noted, graft is not quite the same as in the West. One may steal for the purposes of enhancing comfort but not for gaining wealth. And the official who takes graft must give value to the source. Pure graft, meaning without *quid pro quo,* is against national custom.

I have been impressed in these last days with the confidence of the Thais in relation to Laos, Cambodia and Vietnam. The latter are learning the great truth — Marxian, oddly enough — that socialism, with its enormous demands upon administrative talent, belongs to a later stage of economic development. So, for now, in these countries, there is deprivation and despair. In Thailand, in contrast, there is great and attractive vitality. This leads to a quite wonderful paradox: the strategic theorists who gave us the domino doctrine were, in fact, the captives of the Communist vision. They took for granted that the Communist example would be persuasive. With a little military push, the dominoes would fall. Instead the dominoes are leaning slightly in. The Thai (and Malayan and Singaporean) examples are a threat to Indochina.

At lunch Kukrit spoke at length of his visit to China in 1975. Mao, by then, was unable to speak. An interpreter, half Chinese and half American, read his lips and then checked the meaning with him in Chinese before rendering him into English. Mao nonetheless was exceptionally amusing and jovial withal and told Kukrit that China would not follow the Russian example of imposing its system and influence on reluctant neighbors. This would be not so much out of kindness as because it was impractical.

October 5 — Thursday

Two hours before required check-in time and three hours before the flight was scheduled to go, we left for the airport, a half hour's driving time away. The traffic requires such a margin. It took us half an hour. Kukrit accompanied us out, arranged our passage through the various airport functions, and we then had a couple of hours to review our respective achievements of recent years in a suitably self-laudatory way. A unique companion.

Yesterday I was reminded of the ceremony in ancient times governing the royal succession in Siam. All male claimants seated themselves, lotus position facing inward, around an area spread evenly with honey. Each then drew back his member, as it is decently called; the one who, on release, killed the most flies won the succession to the throne. This, incidentally, was how the capital got its name.

October 7 — Saturday

A day lost from this chronicle. We arrived a bit late, night before last, at Narita Airport, to a ceremony comparable with anything since Admiral M. Calbraith Perry. Television cameras, a dozen news photographers, publishing company executives, television men, a press conference and then champagne and a small banquet, all at the airport. More than Houghton Mifflin would do for John Milton. It was toward midnight be-

fore we left for Tokyo. There being little traffic, it took only
an hour and a half.

"Why is Narita so far from town?"

"Everyone is asking."

Yesterday began with a visit with the executives of TBS —
Tokyo Broadcasting System — who, with the assistance of the
exceptionally enlightened Kirin Beer Company, broadcast my
series *The Age of Uncertainty* to the Japanese viewing public.
Then there was a radio interview — questions in English repeated by the interrogator in Japanese, answers and translation.
It was both faster and funnier than one could have expected.
Then I had some lunch with book and television executives
and an hour-long television discussion with Masayoshi Ohira,
former Foreign Minister, now Secretary General of the Liberal
Democrats and a candidate for Prime Minister. [This office he
has since achieved.] Ohira, an acquaintance from past times,
masks a very sharp mind with a deceptively bland manner. Our
discussion was of industrial development. I asked if the days of
aggressive, single-minded economic growth might not now be
in the past in Japan. He thought so; in the future more attention will be given to the quality and tranquillity of life.

After Ohira came a long interview with *Mainichi Shimbun,*
one of the three biggest newspapers in Japan, an interview
which covered the world and several pages this morning and
is to be continued tomorrow. Then at the end came a dinner
with TBS executives, Ambassador Mike Mansfield and his wife
Maureen and the Shigeto Tsurus. Tsuru is the most distinguished of Japanese economists — a Harvard graduate and
Ph.D. and a friend of mine for some forty years. Mike looked
rested and relaxed, which he attributed to being away from
Washington. We discussed the Japanese penchant for pessimism
and insecurity. All questions half assume the inevitability of
disaster. Someone explained that it is ethnic: Americans expect
to prosper; Japanese regard good fortune as highly unnatural.
There was discussion as to why when the Swiss franc gains and

the Swiss balance of payments booms, all is attributed to the
virtue of the Swiss — solid, reliable folk. When the same happens to the yen and the Japanese balance of payments, as now,
all agree that the Japanese are at fault. The point was not
resolved.

At any given time Washington economists have a fantasy that
gains substance from each telling the others of its depth of
truth. The current one is that Germany and Japan should stimulate their economies more aggressively. They would then have
more inflation, and the dollar would be stronger because the
mark and the yen would be weaker. By this formula we would
remedy our errors by getting other countries to imitate them.
The point came up often during the day, for a huge U.S. trade
mission is in town with carefully coordinated explanations of
the inadequacy of American economic policy. I commented
with unnatural grace and restraint.

This morning we caught the "bullet train" to Atami on the
Izu Peninsula and the peace of Hotel Torikyo, meaning Land
of Peach and Pear. The comfort is perfect. Chaste rooms; beds
prepared on the floor; our own garden; the sea below and beyond. Dinner was with the Tsurus and former Prime Minister
Takeo Miki and his wife. Miki became Prime Minister to give
the Liberal Democrats a new image of honesty, and he is still
regarded as dangerously upright. The subject of discussion: all
know the dark prospect of life, that being nuclear catastrophe.
What are the bright prospects? Probably the tendency for the
socialist and nonsocialist worlds to converge, to the dismay of
the professional ideologues in both. The underlying and less
agreeable tendency is that public and private activity in both
becomes increasingly organized and bureaucratic. The organization man is much the same everywhere. All countries de-
nounce bureaucrats but have no substitute.

The most intractable problem, my friends urged, is the great
and growing difference in wealth between the rich countries
and the poor. A transfer of capital from the rich to the poor

has value but is not a cure. Tsuru, one of the world's leading philosophical Marxists, shares my doubts about socialism for the poor lands and its tendency to load the greatest administrative burden on the countries least able to bear it.

Population pressure is also intractable. One discusses all possibilities without optimism.

October 8 — Sunday

After a parboiling in the local hot spring — there is a small pool in our own bathroom — I worked through the morning on my lectures. I had arrived with good drafts, but quite a bit of cross-cultural and last-minute information has always to be added. Eight thousand people have applied for around nine hundred seats. Admission is by lottery; cassettes will be sold. Harvard undergraduates were never so discriminating. At lunch Tsuru gave me the results of a poll recently conducted by *Mainichi*. What Japanese and what non-Japanese would you prefer for prime minister of Japan? Living or dead could be chosen. I nosed out Einstein in the external competition, an indication of the power of economics and television as compared with mere theoretical physics. Ronald Reagan was nowhere.

This afternoon we toured the local fishing villages and seacoast — a landscape of breathtaking views.

October 10 — Tuesday

The bullet train back to Tokyo yesterday rolled into Atami station on the precise minute and, I would judge, the precise second. A large crowd waited, but there was room in the spacious blue-gray rooms for all. A most pleasant way to travel that other countries should adopt.

In Tokyo I went to the Press Club for a lively luncheon, a couple of hundred attending. My thesis for the day: in the last century and early on in this one, nations sacrificed internal stability and accepted heavy unemployment in order to maintain stable international exchanges. Now the reverse: internal stability and external instability. However, some countries, including the United States, are having some success in getting instability in both areas. The circumstances and the absence of alternatives dictate the action — restraints on corporate, trade union and other organized power. The questions were sharp and to the point. One man did ask if I thought Japan was in a recession. I told him this was hard to perceive.

After lunch we went to TBS-Britannica, my publisher, we being a caravan of three cars, one of which carries photographers and one assistants and logistical support. At TBS-Britannica, the whole staff, to the number of a hundred or so, was down on the street to cheer us in. Then, after visiting the president, all hands assembled in the main offices for speeches, toasts and the presentation of a stunningly beautiful *No* statuette, which filled me with pleasure.

Toward evening prizes were given in a downtown hall in an essay contest on the subject "Japan in the Age of Uncertainty." The winner had dealt with the uncertainties of health from firsthand knowledge. He was a paraplegic in a wheelchair. The second prize went to a woman well into her seventies. She didn't seem disturbed when referred to as "this old woman." In Japan age is still good. I followed with my lecture to the happy few. It was unduly complicated, and I outpaced my interpreters. A very poor performance.

October 11 — Wednesday

I continue a day behind. Yesterday morning we visited two bookstores of enormous size, one of them Yaesu Book Center, said to be the largest in the world. Each is of several stories, each floor two or three times Doubleday Fifth Avenue and packed not only with books but also with customers. A juvenile section at Kinokuniya, the first store, was filled shoulder to shoulder with youngsters aged eight to ten, all diligently reading free. The owner, aging, articulate, highly amusing and a local political mogul, Moichi Tanabe, led our procession through the store. Pointing to the numerous signs of welcome overhead, he assured me that my honor was fleeting. All would be taken down tomorrow. We then went to the handsome headquarters of Soka Gakkai, the large Buddhist religious group (some 13 million members), to meet Daisaku Ikeda, its leader. With his followers he is a force of considerable power in the East. A hundred or so attractive women were outside and in the halls to welcome us. Our meeting was in the footsteps of such diverse visitors and visited as Chou En-lai, Kissinger, Kosygin, Kurt Waldheim and Arnold Toynbee, not in order of importance. First we had questions and answers over sherry and smoked salmon and then over an elegant Japanese luncheon featuring large and small shrimp cooked in the hollow square of a large rectangular table, a variety of fish and vegetables, beer, saki and Moselle and much else that I failed to notice. The first question was whether I believed in a life after life. I thought I might paraphrase Mencken's answer and say: "No, but if I am wrong, I will walk up to God in a manly way and say, 'Sir, I made an honest mistake.' " Instead I fell back on a more general commitment to continuity and added that I would soon have an opportunity for firsthand investigation. The next question was easier.

"What is happiness?"

"The respect of good men and the affection of lovely women."

We then passed on to practical matters — China, Russia, control of atomic weapons, help to the poor lands, Japan's special responsibilities there. When asked, I asked back; there was little on which to disagree. Ikeda was especially concerned about U.S. relations with China. I urged, as usual, that history would see the hostility between us in the twenty-five years following the Chinese Revolution as an aberration and that even our more compulsive anti-Communists — Henry Jackson, the Committee on the Present Danger — have now made the Chinese Communists honorary non-Bolshevists. The greater danger was from our excessively self-instructed foreign policy wizards — the Milli-Metternichs — who wish to play games as between the Soviets and the Chinese. Useless and dangerous. Ikeda agreed. We also agreed that Nixon and Kissinger deserved a bow for breaking through the crust of opposition on relations with the Soviets and the Chinese. Ikeda then urged me to a deeper Buddhist commitment to life's never-ending stream, to its guidance into a calm and pacific course. This I promised.

In the evening there was a seminar with fifty or sixty Japanese economists and business leaders — "The Theatre of Ideas." A current of anxiety about Japan's future ran, as usual, through the questions. Growth cannot continue at recent rates; what happens when it slows down; what, in effect, are the social tensions when a fixed, not an increasing product must be shared? And what — a familiar note — of the intruding presence and the higher costs of government? One or two speakers cited the inefficiency of Japanese government services, the railroads being an example. In any other country the Japanese railroad system would seem a thing of unbelievable efficiency.

I was pressed on the tax revolt in the United States. I returned the thought that it is a revolt of the affluent, who pay the larger share of the taxes, against the poor, who, in education, police, urban and health services, recreation and welfare, are the most dependent on public services and support. And there is the magic belief of some that taxes and expenditures can be cut and services kept the same. This, I ventured, might not be so.

October 12 — Thursday

Yesterday we called on Mike Mansfield in the new — since my last visit — sparse skyscraper that maintains the American presence. Mike was as agreeable and lovable as ever. We talked of economics. He thinks we will have to come to wage and price controls, as does Jim Rowe, who is visiting Tokyo and the Mansfields. Rowe, a major Washington lawyer, was one of the early, anonymous assistants to F.D.R.

From the Embassy I went to address the International Press Club, an audience of Japanese and overseas correspondents presided over by the local CBS man. There being no problem of translation, it was considerably easier than appearances before Japanese groups, where, because of the interpreter, all humor, all obliquity and especially all irony are a risk.

Then we boarded the subway to the main Tokyo railroad station, guides, photographers, *et al.* accompanying. By car in Tokyo traffic, (as in Bangkok) you either allow an excessive margin of time — an hour or so a mile — or miss the train. Our particular subway was fast, clean and uncrowded. At the station we waited while a crew, five or six to a car, cleaned the train. Departure for Osaka was four or five minutes behind time; we couldn't leave until everything was spotless.

October 13 — Friday

Yesterday morning we drove through the urban jungle of Osaka for several miles, arriving eventually at the headquarters of Matsushita Electric, the largest maker of color television sets in Japan. We then had an hour or so of discussion with Konosuke Matsushita, the eighty-three-year-old founder of the company and of a school of politics and government he is establishing near Tokyo. I was pressed on the experience of the Kennedy School of Government in Cambridge and told him we found politicians quite educable — new congressmen go to the school after election and now also new mayors. It is important to have an admixture of young students with senior and experienced civil servants. Our host thought it possible and important to inculcate an improved level of ethics and a stronger sense of the philosophy of public service. From conversation we advanced to an excellent lunch; I am full of admiration for the chefs employed by Japanese corporations. Then we went on a tour of the plant. Machines stamped some components into the sets; some hundreds of exceptionally tidy, blue-uniformed girls stamped more. Gray-clad young men ran the heavier machines. A five-minute break came, and all did calisthenics. Large overhead signs exhorted everyone to an extra 10 percent. Because of the appreciation of the yen and our restrictions on imports, some production is being shifted to a plant near Chicago. Productivity there is lower "but improving." The American work force does not lend itself to calisthenics.

At around 6:30 P.M., I gave a lecture in a vast hall where the exceptionally diligent had been waiting since three, and it was much better than in Tokyo. If something must be translated into Japanese, the pace must be very slow and the ideas in straightforward form. After dinner with numerous booksellers and book wholesalers we went on by car to Kyoto. We are installed in the Japanese-style pavilion of the Miyako Hotel with four or five rooms, two baths and a bottle of Glenfiddich. Somehow the news spread that I like malt Scotch, which I do much better than it likes me. So a bottle is always waiting. My wife thinks Fitzgerald, Hemingway and Faulkner all went wrong from visiting Japan.

October 14 — Saturday

Yesterday our convoy went to Nara, the ancient eighth-century capital of feudal Japan, or such part as could then be considered subject to central rule. We surveyed a Buddha of giant proportions, which, like one in Hangchow, China, and a third which reclines in Bangkok, is described as the largest in the world. An unseemly competition. The temples were interesting, some of the gardens lovely and the visiting tourists most interesting of all. The latter were uniformed schoolchildren of identical age groupings, all brought thither by bus. They were eager, noisy, disciplined and clean. Discipline suffered severely when I was recognized. Television as ever.

We lunched grandly at a local hotel and made our way back to Kyoto, the journey, as seems always so in Japan, through an endless industrial and residential wasteland. Traffic was just below total stasis. Our guide for the day was an attractive former executive at my publisher's, whom I judge to be as reliable as she is charming. We agreed that Japan combines a commitment to Japanese manners with an obsession with American style. She held that discipline among the young is still fairly strong. In elementary and high schools the desire to get into university remains a major incentive. (In Nara there is a fence where, after writing out one's prayers, one twists them onto the wire. A very large number ask success on examinations and consequent passage into university.) However, sex has been discovered in the high schools; teenage pregnancy is increasing; the pill is not a sufficient antidote; and while drugs are not a problem, the sniffing of paint-thinning fluid is popular. In the universities idleness is fashionable and, as in the United States and especially Europe, extensively identified with reform.

We eventually arrived back in Kyoto and went to an ancient and beautiful house, a several-hundred-year-old place of aristocratic rest and relaxation and a literal dream of flower arrangements, wall paintings and gardens. The tea ceremony was by a *taiyu,* a woman socially several steps above a *geisha.* In past times a *taiyu* might have been the mistress of some major feudatory but with the culture and personal dignity that allowed her to reject a lover of any rank whom she found inadequate in manners or intellect. I liked ours very much. She was magnificently dressed, and two small girls, equally well accoutered, assisted her in the ceremony.

Then we had dinner at Kitcho, described as the best restaurant in Japan. There was no evidence to the contrary. *Geishas* sat at our elbows, anticipated our choice of victuals and then played and danced for us as the meal came to an end.

This morning I had a discussion of various publication matters. At lunch a cook doing the beef in front of us whipped out a book to be autographed. We were fifteen minutes early at the Kyoto station. Before our train arrived — precisely on time — three other express trains arrived and departed on the same track in the same direction. Several locals moved on more distant tracks.

The entire distance from Kyoto to Tokyo, 310 miles by rail, is now urbanized, the hills above the railroad tunnels and a few odd stretches partially apart. Rice lands (and vegetables and some tea) are interspersed, lot by lot, with single family houses, industrial plants, storage yards, junk yards and parking lots. On not everything are the Japanese careful; in a land-scarce country there is a prodigal waste of land.

October 15 — Sunday

After we arrived in Tokyo last night, Tsuru and I had a long recorded conversation of much interest to me and perhaps also to the audience. He thinks the overvaluation of the yen (meaning that it buys more when converted into dollars or other currencies than it does in Japan) will soon bring a shrinkage in Japanese exports, a reduction in the present surplus of exports over imports and some decline in the rate of growth. This means that Japan will not reach the 7 percent expansion in Gross National Product "agreed on" [sic] at a recent meeting of chiefs of state in Bonn. He thinks further that the Japanese economy is maturing rapidly and encountering, in consequence, competition from lower-wage countries. Steel production is beginning to move from Japan to Korea, shipbuilding more so. We agreed that setting growth targets by international conference agreement is a sublime absurdity. It implies far more capacity for economic management than exists. Economic advisers persuade politicians (as they earlier have persuaded themselves) that such goals have meaning. Negotiation then proceeds which is without content or common sense. Neither the public nor the politicians know enough to expose the fraud.

After dinner we went to a small nightclub in the Ginza, the first such resort I've visited in some decades. It is a wholesome place given to music and alcohol "where you can bring your wife." We were joined by some articulate and very wholesome-looking young women in conservative western dress. One singled me out and pressed me on my books. I asked her the source of her excellent and idiomatic English.

"Mostly in bed, but I also studied it in school."

This morning — my birthday — we went to the top of the Okura Hotel for a massive party combining intricate and graceful Japanese ceremony with the singing of "Happy Birthday" and "Auld Lang Syne" and a cake that I cut with a Samurai sword. Then came the long drive out to Narita. Various security procedures preceded another large party. We are now taking off in heavy rainfall, and the pilot has promised rough weather ahead.

October 15 — Sunday (again)

Today, because of the time zones, was also and again October 15. Two birthdays at an age when having one is clinically depressing. After Tokyo, San Francisco looked spacious, leisurely and very dirty.

October 16 — Monday

We slept last night at the Stanford Court, a worthy hotel on Nob Hill, far from inexpensive. This morning the Boston plane is filled with doctors returning from a convention of the American College of Surgeons in San Francisco. Until a short time ago all in adjacent seats and across the aisles thought themselves in a convivial mood. In point of fact, they were obscenely noisy as they shouted greetings and insults to one another across the plane. Eventually I told them with great authority that they were to be quiet. They obeyed. All are now silent, even glum. The chief flight attendant, a tall, personable, slender young woman, just sat down beside me to render warm thanks.

Little more is expected to happen on this journey.

III

The Arts and…

20

Evelyn Waugh

"WE WERE JOINED at lunch by an ungainly and deeply garrulous American who called himself an economist and spoke admiringly of my books and piously of the prospect for socialism and similar depravities in the United States. Eventually I was forced to tell him it was a subject in which I had no interest whatever."

I never met Evelyn Waugh so this entry cannot be found in his diaries.[1] Everything else was calculated to repel me. Except for Randolph Churchill, a couple of the Mitford sisters, a writer or two and one economist, the late Sir Roy Harrod of Oxford, I know or knew none of the people mentioned. Nor did I feel deprived. I learned some years ago that alcohol and barbituates, either separately or in modest combination, interfere with writing and, increasingly as you grow older, make you sick. Accordingly, there was no surprise in Waugh's endless chronicle of inebriation, drugs and their consequences. As an American liberal with impeccable credentials in the faith, I don't like slighting references to Jews, "Negroes" and manual workers, and I don't believe that ever in my life I have actually heard anyone refer to working people as "the lower classes." In the diaries such references abound. My early religious education, if such it can be called, was extensively concerned with the li-

[1] *The Diaries of Evelyn Waugh*, edited by Michael Davie (London: Weidenfeld and Nicolson, 1976).

turgical excrescences of the Catholic Church, as we were required to regard them. The virtue of these, at least in their traditional form, was one of the few things in life which Waugh held in unwavering reverence, and they get much attention in his diaries, although considerably less than liquor does.

Yet I read compulsively and from the beginning of the diaries to the end. So, I think, will thousands of others whose knowledge of the people and scene is no greater than mine.

On two matters, his travels and his World War II experiences, Waugh is a serious diarist. He writes well and vividly and gets perceptively into his subject. In consequence, although it is hard to believe that this was his intention, he makes a genuine contribution to the understanding of modern warfare. All who have had even the most marginal association with martial enterprise know the sheer awfulness of having to deal with drunks and incompetents suddenly released from the penalties and constraints of civilian existence and endowed with military swagger and some power. All know the damage they do, the foul-ups they perpetrate and how this behavior and its consequences are diminished in both the histories and the unwritten (as distinct from the American Legion) recollections. This is especially so on the side that won. Waugh's postwar trilogy — *Men at Arms, Officers and Gentlemen, The End of the Battle* — tells this history, but, as with *Catch-22,* one attributes some or much to invention. The diaries, telling of training in Britain, the expedition to Dakar, staging in Egypt, fighting in Greece and his mission with Randolph Churchill to Tito and the partisans, must be the most sustained account of military fecklessness, incoherence, cowardice and intoxication of World War II and maybe any war. There was, no doubt, a more responsible side; Waugh was not one to struggle for even-handed judgment. But anyone with military experience will know how great was the balance to be redressed.

Social purpose, however, is not Waugh's claim to accomplishment — the thought alone is slightly bizarre. His claim, even when telling of aristocratic nonentities, drugs or drunkenness, is in the way he tells it. Many have said it before: there was not in our time, perhaps in our century, such a master of the craft. As a test, one need only open the diaries at random. There is, for example, the name of Ruth Blezard. I have never heard of her, but I learn with immediate delight that she "is a fat woman with ugly hair and a puzzled expression in her eyes" and that her father "is paralysed through evil living" and "her mother [is] a fool." So much for Ruth and her parents. Next day some boys visiting in the Waugh household were "allowed to listen to a concert on the wireless because [the father of one] ... was singing. It confirmed me in the detestation of the invention." Two days later comes one of the best entries of all: "I think nothing happened." Even the schoolboy passages show signs of the quick, flashing stroke that disposes of everything and everyone.

Somewhere at Harvard, and perhaps also at Cambridge and Oxford, young scholars are dissecting Waugh's sentences, ascertaining, probably with the aid of a computer, the nature of their magic. I await the result without interest. Waugh's art is safely beyond the reach of research.

It leads me to a harsh word in conclusion for the editor of the diaries. Though, I'm sure, a scholar and a gentleman, Mr. Davie, either on his own motion or (more rarely) at the behest of the Waugh estate, felt obliged to delete "libellous" and "offensive" passages as well as other entries that might have caused unnecessary distress. In justification, he cites Waugh's "taste for exaggeration and fantasy" and the resulting danger of misrepresentation. This is trite, blithering nonsense! Is British justice dead? Would any British judge award damages to a plaintiff, otherwise only a titled nullity, who had achieved the supreme distinction of being misrepresented by Evelyn Waugh? The judge, if he has any right to be on the bench, would hold that the plaintiff should pay for being so singled out, so honored. And as to the exaggeration, fantasy and misrepresentation leading to "unnecessary distress," no decently literate person in the English-speaking world can surely believe that because Waugh said something it was true.

One cannot read these diaries without sensing that Waugh's firmest intention was to be nasty, even vicious. No one, I assert, had the right to frustrate so clear a purpose, and certainly not after the author is dead.

21

Anthony Trollope

IN THE MID-NINETEEN-SEVENTIES, the novel came into its own as a medium for highly topical political and behavioral comment, Spiro Agnew, John Lindsay, John Ehrlichman and Ms. Elizabeth Ray, the nontyping friend of former Congressman Wayne Hays, being the principal protagonists. John Ehrlichman, perhaps as an exercise in some obscure form of retributive justice, suffered here from guilt by association. As a writer, he had an excellent sense of scene, could sustain suspense and seemed capable of a substantial exercise of imagination. But his own character, personality and service to Richard Nixon rather than his writing got reviewed by some critics. That for Ehrlichman was not good. One may also be a little unfair to Ms. Ray in associating her with Mr. Agnew. Her book was five times as good as his, for it was only one-fifth as long. Reviewing Mr. Agnew's book, I said that it had "one hell of a plot." I meant it as one would say, "he had one hell of a cold." The publisher seized, nonetheless, on the ambiguity and used the quotation in his advertising for many weeks. I thought of sending a letter of explanation to the Federal Trade Commission.

However, my purpose here is not to discuss the work of these former and soon-to-be-forgotten public servants, if one may speak loosely of some, but to recommend as an alternative the novels of another and earlier public figure, the British postal official Anthony Trollope. I turned to him between Agnew and

Ehrlichman, and not at all by accident. I turn to him constantly when I need some improvement over the present. That time it was to *The Last Chronicle of Barset,* which I consider the best of Trollope's works, as earlier, after rereading *Barchester Towers, Doctor Thorne, The Eustace Diamonds, Orley Farm* and *Phineas Finn,* I had considered each of *them* the best. There is, however, a higher recommendation for *The Last Chronicle.* Of it Trollope said: "Taking it as a whole, I regard this as the best novel I have written."

The Last Chronicle resembles the work of the more modern officials and acolytes in two significant respects. It was written, like everything Trollope did, with an eye to the money. It earned £3000, which, assuming a three-fold deterioration of the pound in relation to the dollar since 1867 and a roughly four-fold deterioration of the dollar, would round out at about $35,000. (Trollope estimated his lifetime earnings from writing at £70,000 or, by the same calculation, around $850,000. The income tax was nominal.) Then as now, though possibly with less reason, obsession with cash was thought inconsistent with the artistic spirit, and Trollope's detailed accounting of his revenues in his *Autobiography* has always been thought one of the things that put him under a cloud for fifty years or so after his death. It's likely that his work habits had more to do with it. "While I was in Egypt, I finished *Doctor Thorne,* and on the following day began *The Bertrams.*" All writers want it thought that a long respite is required between their orgies of effort. Trollope was also merciless in arguing that alcohol is as inimical to writing as to shoemaking. This few wish to believe. Finally, he reminded his colleagues, as I have urged elsewhere, that waiting for inspiration by a writer, the most common of all excuses for indolence, is as absurd as by a "tallow-chandler for the divine moment of melting." All of these beliefs were in conflict with the union rules.

Along with money, Trollope, like the more recent novelists, greatly loved politics. This affection was most strongly manifested in the parliamentary novels, which tell, along with much else, of Plantagenet Palliser's crusade, almost exactly a century ahead of success, to provide the English with a decimal currency. *The Last Chronicle,* however, is also a political novel, it being about clerical politics, and no one should suppose that the most subtle and ruthless practice of this art is to be found in legislatures or bureaucracies. In the early days of his presidency Gerald Ford had no fewer than three Harvard professors — Henry Kissinger, Daniel Patrick Moynihan and John Dunlop — in his Cabinet and, for good measure, another former faculty member, James Schlesinger. Even Mr. Ford must have sensed the unique subtlety, skill and malignity with which these scholars, drawing on their Harvard faculty experience, operated in Washington. And let all recall the exquisite grace with which Kissinger disposed of both Moynihan and Schlesinger.

The opposing clerical factions in Barset were equally accomplished. One was headed by Archdeacon Grantly, the son of the old Bishop, and a very tenacious man with a grudge. The other was captained by Mrs. Proudie, the wife of the present Bishop, an antagonist before whom Bella Abzug would have quailed and surrendered. A seeming curiosity of the political struggle in Barset was that no one quite knew what it was about. The distribution of power was accepted by everyone but Mrs. Proudie. There was patronage, but its dispensation was so crass and flagrant, and also so completely regulated by custom, that there was almost nothing to quarrel over. The Grantly faction was High Church, Mrs. Proudie relentlessly Low, but this conflict never went deeper than approval or disapproval of foxes as distinct from fox-hunting. At Harvard, in years past, I used to think that the major political issue had to do not with appointments, research funds or ideas but with prestige. Did it belong, as a matter of right, to the president of the university, in my time a dutiful, mannered, well-scrubbed gentleman of no scholarly pretense, and those of equally fine bearing who up-

held his hand and occupied the offices of presumed distinction
in University Hall? Or was prestige properly the possession of
the pushy, raucous, less couth figures who found their distinction in their science or subject matter or the world at large?
The political struggle in Barset was the same. Maybe it was the
genius of Trollope that he saw that politics does not need
issues, only competition for esteem.

On one feature, however, no contrivance can produce a parallel between Trollope and the recent political novelists. His
characters are unforgettable where those of the modern writers
may be as forgettable as any in history. There are three superb
people in *The Last Chronicle*. One is the Reverend Septimus
Harding, who, in the remunerative sinecure of Warden of Hiram's Hospital, had, years earlier, launched Trollope in Barchester in *The Warden*. The brief chapters on his decline and
death in this novel are wonderfully affecting. Then there is
Mrs. Proudie, whom Trollope later described as "a tyrant, a
bully, a would-be priestess, a very vulgar woman, and one who
would send headlong to the nethermost pit all who disagreed
with her; but . . . conscientious, by no means a hypocrite, really
believing in the brimstone which she threatened, and anxious
to save the souls around her from its horrors." She too dies in
The Last Chronicle, suddenly of a heart attack after subjecting
the Bishop to terrible humiliation in pursuit of justice and
virtue as she sees them. One is at first relieved to see her go but
then discovers with sorrow how empty the world has become.
That is because Trollope felt the same way. While working on
the novel one day in the Atheneum Club in London, he over-
heard two clerygmen discussing it. (In keeping with contemporary practice, the novel was being serialized as written.) They
were giving Mrs. Proudie the business and in very severe terms.
Trollope rose, advanced on them and said that, in light of their
disapproval, he would promptly kill her off. He delighted in
their shock and embarrassment and then almost immediately
regretted his promise. This regret he makes the reader share.

But his supreme achievement, perhaps his greatest ever, is the Reverend Josiah Crawley, the perpetual curate (meaning the pathetically underpaid minister) of the poor parish of Hogglestock. Crawley, a most learned man, is a masterwork of psychological conflict. He believes himself to be both greatly superior and deeply inferior to other men. He loves his family and insists on their deprivation and abuse. He demands kindness and sympathy from others and profoundly resents it when they are displayed. On a scale of personal rectitude which would put the Nixon men around zero, the average honest citizen around seventy, Crawley would come in at maybe a hundred and fifty. That is because he is not only relentless and unbending as regards himself, but he is impelled to make everyone else the same. The plot concerns a check which, *in extremis,* he passes on and for the possession of which he cannot account. He comes to believe, because he is too honest to deny the possibility, that in desperation he must have stolen it. So there is a final climactic conflict between Josiah Crawley's view of himself as a paragon of integrity and the seeming fact that he is a thief. With all else, Crawley is given to psychotically depressive episodes and extreme absentmindedness.

As with other of Trollope's characters, Crawley is also incapable of conversation, only of speeches. At the end of the novel, cleared of the charge of theft, he is offered a much more remunerative living.

"And you will accept it — of course?" his lovely daughter asks.

"I know not that, my dear. The acceptance of a cure of souls is a thing not to be decided on in a moment — as is the colour of a garment or the shape of a toy. Nor would I condescend to take this thing from the archdeacon's hands . . ." The speech continues.

The recent political novelists are also great on speeches as a substitute for conversation. Still, Trollope's ending is a document on man's moral progress in the last century. The ending is happy because Crawley proves to be corruptible. He leaves the poor people of Hogglestock and Hoggle End who are devoted to him and where, clearly, his clerical obligation lies, and goes to the higher remuneration and much easier life of suburban St. Ewold. Everyone of importance in Barset is overjoyed. Honesty and duty have yielded to reason and moderation. No recent author has had such a thought.

22

Writing and Typing

NINE OR TEN YEARS AGO, when I was spending a couple of terms at Trinity College, Cambridge, I received a proposal of more than usual interest from the University of California. It was that I take a leave from Harvard and accept a visiting chair in rhetoric at Berkeley. They assured me that rhetoric was a traditional and not, as one would naturally suppose, a pejorative title. My task would be to hold seminars with the young on what I had learned about writing in general and on technical matters in particular.

I was attracted by the idea. I had spent decades attempting to teach the young about economics, and the practical consequences were not reassuring. When I entered the field in the early nineteen-thirties, it was generally known that the modern economy could suffer a serious depression and that it could have a serious inflation. In the ensuing forty years my teaching had principally advanced to the point of telling that it was possible to have both at the same time. This was soon to be associated with the belief of William Simon and Alan Greenspan, the guiding hands of Richard Nixon and Gerald Ford, that progress in this field is measured by the speed of the return to the ideas of the eighteenth century. A subject in which it can be believed that you go ahead by going back has many problems for a teacher. Things are better now. Mr. Carter's economists do not believe in going back. But, as I've elsewhere urged, they are caught in a delicate balance between their fear of inflation and unemployment and their fear of doing anything about them. It is hard to conclude that economics is a productive intellectual and pedagogical investment.

Then I began to consider what I could tell about writing. My experience was certainly ample. I had been initiated by two inspired professors in Canada, O. J. Stevenson and E. C. McLean. They were men who deeply loved their craft and who were willing to spend endless hours with a student, however obscure his talent. I had been an editor of *Fortune,* which in my day meant mostly being a writer. Editor was thought a more distinguished title, and it justified more pay. Both as an editor proper and as a writer, I had had the close attention of Henry Robinson Luce. Harry Luce is in danger of being remembered only for his political judgment, which left much to be desired; he found unblemished merit in John Foster Dulles, Robert A. Taft and Chiang Kai-shek. But more important, he was an acute businessman and a truly brilliant editor. One proof is that while Time, Inc. publications have become politically more predictable since he departed, they have become infinitely less amusing.

Finally, as I reflected on my qualifications, there was the amount of my life that I have spent at a typewriter. Nominally I have been a teacher. In practice I have been a writer — as generations of Harvard students have suspected. Faced with the choice of spending time on the unpublished scholarship of a graduate student or the unpublished work of Galbraith, I have rarely hesitated. Superficially at least, I was well qualified for that California chair.

There was, however, a major difficulty. It was that I could tell everything I knew about writing in approximately half an hour. For the rest of the term I would have nothing to say except as I could invite discussion, this being the last resort of the distraught academic mind. I could use up a few hours telling how a writer should deal with publishers. This is a field of study in which I especially rejoice. All authors should seek to establish a relationship of warmth, affection and mutual mistrust with their publishers in the hope that the uncertainty will add, however marginally, to compensation. But instruction on how to deal with publishers and how to bear up under the inevitable defeat would be for a very advanced course. It is not the sort of thing that the average beginning writer at Berkeley would find immediately practical.

So I returned to the few things that I could teach. The first lesson would have had to do with the all-important issue of inspiration. All writers know that on some golden mornings they are touched by the wand; they are on intimate terms with poetry and cosmic truth. I have experienced those moments myself. Their lesson is simple; they are a total illusion. And the danger in the illusion is that you will wait for them. Such is the horror of having to face the typewriter that you will spend all your time waiting. I am persuaded that, hangovers apart, most writers, like most other artisans, are about as good one day as the next (a point that Trollope made). The seeming difference is the result of euphoria, alcohol or imagination. All this means that one had better go to his or her typewriter every morning and stay there regardless of the result. It will be much the same.

All professions have their own way of justifying laziness. Harvard professors are deeply impressed by the jeweled fragility of their minds. Like the thinnest metal, these are subject terribly to fatigue. More than six hours of teaching a week is fatal — and an impairment of academic freedom. So, at any given moment, the average professor is resting his mind in preparation for the next orgiastic act of insight or revelation. Writers, by the same token, do nothing because they are waiting for inspiration.

In my own case there are days when the result is so bad that no fewer than five revisions are required. However, when I'm greatly inspired, only four are needed before, as I've often said, I put in that note of spontaneity which even my meanest critics concede. My advice to those eager students in California would have been, "Don't wait for the golden moment. Things may well be worse."

I would also have warned against the flocking tendency of writers and its use as a cover for idleness. It helps greatly in the avoidance of work to be in the company of others who are also waiting for the golden moment. The best place to write is by yourself because writing then becomes an escape from the terrible boredom of your own personality. It's the reason that for years I've favored Switzerland, where I look at the telephone and yearn to hear it ring.

The question of revision is closely allied with that of inspiration. There may be inspired writers for whom the first draft is just right. But anyone who is not certifiably a Milton had better assume that the first draft is a very primitive thing. The reason is simple: writing is difficult work. Ralph D. Paine, who managed *Fortune* in my time, used to say that anyone who said writing was easy was either a bad writer or an unregenerate liar. Thinking, as Voltaire avowed, is also a very tedious process which men or women will do anything to avoid. So all first drafts are deeply flawed by the need to combine composition with thought. Each later one is less demanding in this regard; hence the writing can be better. There does come a time when revision is for the sake of change — when one has become so bored with the words that anything that is different looks better. But even then it may *be* better.

For months when I was working on *The Affluent Society,* my title was "The Opulent Society." Eventually I could stand it no longer; the word opulent had a nasty, greasy sound. One day, before starting work, I looked up the synonyms in the dictionary. First to meet my eye was the word "affluent." I had only one worry; that was whether I could possibly sell it to my publisher. All publishers wish to have books called *The Crisis in American Democracy*. The title, to my surprise, was acceptable. Mark Twain once said that the difference between the right word and almost the right word is the difference between lightning and a lightning bug.

Next, I would have stressed a rather old-fashioned idea — brevity — to those students. It was, above all, the lesson of Harry Luce. No one who worked for him ever again escaped the feeling that he was there looking over one's shoulder. In his hand was a pencil; down on each page one could expect, at any moment, a long swishing wiggle accompanied by the comment: "This can go." Invariably it could. It was written to please the author and not the reader. Or to fill in the space. The gains from brevity are obvious; in most efforts to achieve it, the worst and the dullest go. And it is the worst and the dullest that spoil the rest.

I know that brevity is now out of favor. The *New York Review of Books* prides itself on giving its authors as much space as they want and sometimes twice as much as they need. Writing for television, on the other hand, as I've learned in the last few years, is an exercise in relentless condensation. It has left me with the feeling that even brevity can be carried to extremes. But the danger, as I look at some of the newer fashions in writing, is not great.

The next of my injunctions, which I would have imparted with even less hope of success, would have concerned alcohol. Nothing is so pleasant. Nothing is so important for giving the writer a sense of confidence in himself. And nothing so impairs the product. Again there are exceptions: I remember a brilliant writer at *Fortune* for whom I was responsible who could work only with his hat on and after consuming a bottle of Scotch. There were major crises for him in the years immediately after World War II when Scotch was difficult to find. But it is, quite literally, very sobering to reflect on how many good American writers have been destroyed by this solace — by the sauce. Scott

Fitzgerald, Sinclair Lewis, Thomas Wolfe, Ernest Hemingway, William Faulkner — the list goes on and on. Hamish Hamilton, once my English publisher, put the question to James Thurber: "Jim, why is it so many of your great writers have ruined themselves with drink?" Thurber thought long and carefully and finally replied, "It's this way, Jamie. They wrote those novels, which sold very well. They made a lot of money and so they could buy whisky by the case."

Their reputation was universal. A few years before his death, John Steinbeck, an appreciative but not a compulsive drinker, went to Moscow. It was a triumphal tour, and in a letter that he sent me about his hosts, he said: "I found I enjoyed the Soviet hustlers pretty much. There was a kind of youthful honesty about their illicit intentions that was not without charm." I later heard that one night, after a particularly effusive celebration, he decided to return to the hotel on foot. On the way he was overcome by fatigue and the hospitality he had received and sat down on a bench in a small park to rest. A policeman, called a militiaman in Moscow, came along and informed John, who was now asleep, and his companion, who spoke Russian, that the benches could not be occupied at that hour. His companion explained, rightly, that John was a very great American writer and that an exception should be made. The militiaman insisted. The companion explained again and insisted more strongly. Presently a transcendental light came over the policeman's face. He looked at Steinbeck asleep on the bench, inspected his condition more closely, recoiled slightly from the fumes and said, "Oh, oh, Gemingway." Then he took off his cap and tiptoed carefully away.

We are all desperately afraid of sounding like Carrie Nation. I must take the risk. Any writer who wants to do his best against a deadline should stick to Coca-Cola.

Next, I would have wanted to tell my students of a point strongly pressed, if my memory serves, by George Bernard Shaw.

He once said that as he grew older, he became less and less interested in theory, more and more interested in information. The temptation in writing is just the reverse. Nothing is so hard to come by as a new and interesting fact. Nothing is so easy on the feet as a generalization. I now pick up magazines and leaf through them looking for articles that are rich with facts; I don't much care what they are. Evocative and deeply percipient theory I avoid. It leaves me cold unless I am the author of it myself. My advice to all young writers would be to stick to research and reporting with only a minimum of interpretation. And even more this would be my advice to all older writers, particularly to columnists. As one's feet give out, one seeks to have the mind take their place.

Reluctantly, but from a long and terrible experience, I would have urged my class to recognize the grave risks in a resort to humor. It does greatly lighten one's task. I've often wondered who made it impolite to laugh at one's own jokes, for it is one of the major enjoyments in life. And that is the point. Humor is an intensely personal, largely internal thing. What pleases some, including the source, does not please others. One laughs; another says, "Well, I certainly see nothing funny about that." And the second opinion has just as much validity as the first, maybe more. Where humor is concerned, there are no standards — no one can say what is good or bad, although you can be sure that everyone will. Only a very foolish man will use a form of language that is wholly uncertain in its effect. And that is the nature of humor.

There are other reasons for avoiding humor. In our society the solemn person inspires far more trust than the one who laughs. The politician allows himself one joke at the beginning of his speech. A ritual. Then he changes his expression and affects an aspect of morbid solemnity signaling that, after all, he is a totally serious man. Nothing so undermines a point as its association with a wisecrack; the very word is pejorative.

Also, as Art Buchwald has pointed out, we live in an age when it is hard to invent anything that is as funny as everyday life; how could one improve, for example, on the efforts of the great men of television to attribute cosmic significance to the offhand and hilarious way Bert Lance combined professed fiscal conservatism with an unparalleled personal commitment to the deficit financing of John Maynard Keynes? And because the real world is so funny, there is almost nothing you can do, short of labeling a joke a joke, to keep people from taking it seriously. A number of years ago in *Harper's* I invented the theory that socialism in our time was the result of our dangerous addiction to team sports. The ethic of the team is all wrong for free enterprise. Its basic themes are cooperation; team spirit; acceptance of leadership; the belief that the coach is always right. Authoritarianism is sanctified; the individualist is a poor team player, a menace. All this our vulnerable adolescents learn. I announced the formation of an organization to combat this deadly trend and to promote boxing and track instead. I called it the CAI — Crusade for Athletic Individualism. Scores wrote in to *Harper's* asking to join. Or demanding that baseball be exempted. A batter is, after all, on his own. I presented the letters to the Kennedy Library.

Finally, I would have come to a matter of much personal interest, one that is intensely self-serving. It concerns the peculiar pitfalls for the writer who is dealing with presumptively difficult or technical matters. Economics is an example, and within the field of economics the subject of money, with the history of which I have been much concerned, is an especially good case. Any specialist who ventures to write on money with a view to making himself intelligible works under a grave moral hazard. He will be accused of oversimplification. The charge will be made by his fellow professionals, however obtuse or incompetent, and it will have a sympathetic hearing from the layman. That is because no layman really expects to understand about money, inflation or the International Monetary Fund. If he does, he suspects that he is being fooled. Only someone who is decently confusing can be respected.

In the case of economics there are no important propositions that cannot, in fact, be stated in plain language. Qualifications and refinements are numerous and of great technical complexity. These are important for separating the good students from the dolts. But in economics the refinements rarely, if ever, modify the essential and practical point. The writer who seeks to be intelligible needs to be right; he must be challenged if his argument leads to an erroneous conclusion and especially if it leads to the wrong action. But he can safely dismiss the charge that he has made the subject too easy. The truth is not difficult.

Complexity and obscurity, on the other hand, have professional value; they are the academic equivalents of apprenticeship rules in the building trades. They exclude the outsiders, keep down the competition, preserve the image of a privileged or priestly class. The man who makes things clear is a scab. He is criticized less for his clarity than for his treachery.

Additionally, and especially in the social sciences, much unclear writing is based on unclear or incomplete thought. It is possible with safety to be technically obscure about something you haven't thought out. It is impossible to be wholly clear on something you don't understand; clarity exposes flaws in the thought. The person who undertakes to make difficult matters clear is infringing on the sovereign right of numerous economists, sociologists and political scientists to make bad writing the disguise for sloppy, imprecise or incomplete thought. One can understand the resulting anger. Adam Smith, John Stuart Mill and John Maynard Keynes were writers of crystalline clarity most of the time. Marx had great moments, as in *The Communist Manifesto*. Economics owes very little, if anything, to the practitioners of scholarly obscurity. However, if any of my California students had come to me from the learned professions, I would have counseled them that if they wanted to keep the confidence of their colleagues, they should do so by always being complex, obscure and even a trifle vague.

You might say that all this constitutes a meager yield for a lifetime of writing. Or perhaps, as someone once said of Jack Kerouac's prose, not writing but typing.

23

John Bartlow Martin
and Adlai Stevenson

IN THE SUMMER OF 1952, not long after the Democratic Convention in Chicago had nominated Adlai Stevenson, a young *Saturday Evening Post* writer with a reputation for detailed, meticulous and far from abbreviated reporting showed up at the Stevenson campaign headquarters in Springfield, Illinois. He was John Bartlow Martin. Presently he moved from the press office to a literary ghetto that had been established on top of the Springfield Elks Club. Here he became one of Stevenson's best speech writers and certainly his most reliable. His specialty was the short five- or ten-minute airport or whistle-stop address; no one then, and I think no one since, could make a point more succinctly, support it more sharply with evidence and then, surprisingly for him, stop. Martin repeated the performance in 1956, and, as late as 1972, while intense younger colleagues at campaign headquarters were resolving the difficult questions of personnel and policy to be faced by their new president following the election, he was doing the same thing for George McGovern.

During the Kennedy-Johnson years Martin was ambassador to the Dominican Republic, an exceptionally awkward assignment in the aftermath of Trujillo. By then he was already writing a biography of Adlai Stevenson.[1] In this work, now published, he retains the view (which we shared during Stevenson's life) that Stevenson was a greatly misunderstood figure mostly because he relentlessly and imaginatively cultivated such misunderstanding. Martin's biography is really a discovery of Adlai Stevenson. He also brings to his work an encyclopedic and horrifying knowledge of Chicago and Illinois politics and a firsthand knowledge of the people and the scene. He is without malice or meanness, but friendship is never allowed to do down a fact. And he has a nearly uncontrollable delight in detail. He has given it full rein on Stevenson's parents, family, friends, women friends, political allies and foes, and has organized it all into a far more coherent and interesting story than anyone would have thought possible. Some writers take many words to say little; John Bartlow Martin takes many words, but fewer than would be supposed, to say everything.

One learns that Stevenson's father was interested in getting the Zeppelin franchise for the United States; that Adlai's exact salary in law practice was $1450 in 1927 and rose countercyclically to $4847 in 1932; that in the Agricultural Adjustment Administration, before deductions, it was $6500 (more than twice mine as an economist in the same agency at almost the same time); and that, unsullied by prior love, he was married at 4 P.M. on December 1, 1928, in a Presbyterian chapel on the Chicago Gold Coast with a seating capacity of thirty people, the walls of which were covered with greenery. Those who know Martin know that most likely he spent several days trying to track down the particular kind of plant or tree that made the green and was desolate when he failed.

But this essay is not about Martin's history but the man it discovers, for Adlai Stevenson was a compulsive role-player. He presented himself to the world and equally to his friends as he

[1] *Adlai Stevenson of Illinois* (New York: Doubleday, 1976) and *Adlai Stevenson and the World* (New York: Doubleday, 1977).

liked to believe he was. There was a public and a private Stevenson. The real service of Martin's relentless accumulation of fact is in penetrating this extraordinary screen.

Thus, for millions of his admirers, Stevenson was a lonesome, brooding intellectual — a man who made up his mind only after long struggle and then remained uncertain in his view. He loved the image — there all by himself, buffeted by the conflicting evidence and argument. What could he do?

But none of this was so. Stevenson was greatly gregarious. He rejoiced in company, admiration and an audience and hated to be alone. Night or day he rarely was. John F. Kennedy was thought to be much in the arms of his mistresses. He was, I'm persuaded, a Gandhian ascetic as compared with Adlai Stevenson. There were other odd contrasts. Stevenson the brooding intellectual hardly ever read a book. Kennedy, with no similar reputation, was a voracious reader; he accomplished at least twenty books to Stevenson's one.

Stevenson's reputation for indecision was also carefully cultivated. He gave his finest demonstrations of searching self-doubt when he had already made up his mind — as all with a different view eventually discovered. On small matters, such as whether to go out to dinner or a party, he could be endlessly troubled, although in the end he almost always went. On important matters, though he was responsive to opposing views, it was only rarely and very gradually to the point of changing his own.

The thespian Stevenson was at his best on the American political and campaign scene. There he presented himself as a harried and driven figure made to do things that were unnatural to his character or disastrous to his career, of which the most notable case, of course, was his draft by the Democrats in 1952. Later from his campaign plane came a stream of letters to friends, most of them to Alicia Patterson (whom he called Elisha), telling of the indignities to which he was subject, his doubts as to whether it would ever be over. As Martin tells (and as those who were with him knew), Stevenson, after protesting the political show, always acquiesced, and beneath all he greatly liked coming into town to the music of the high school band. There is more than a hint in Martin that, for at least a year before the Chicago convention, he had his thoughts on the nomination. Maybe he saw or sensed that reluctance was precisely the thing that would set him apart from the other contenders. No reluctance in later years kept him from being other than extremely available for the United Nations delegation, State Department employment or the Illinois senatorial nomination.

A greater mystery surrounds Stevenson's economic and political views. Here there is a genuine question as to his beliefs, whether he was a liberal or a conservative. He did have a stalwart commitment to civil liberties in a difficult time. But he had an equally solid commitment to the American class structure — specifically to a privileged and affluent elite of which he was happily a member. In 1952, he had grave doubts about unions. And he regarded Keynesian economics, to the extent that he understood it, as a transparently spurious alibi for a sloppy management of public finances. He had given no attention to any matter seriously related to economic policy since his undergraduate days at Princeton. There he had been exposed to a rigorously primitive version of classical economic theory.

Martin tells of Willard Wirtz, later Secretary of Labor under Kennedy, and Carl McGowan, now a judge of the Circuit Court of Appeals, struggling with the text of Stevenson's Labor Day speech in Detroit in 1952. So, for some troubled days, did I. Our problem was that Stevenson was basically in sympathy with the Taft-Hartley Act, then for the unions the symbol of all capitalist repression. There was no chance that he would urge repeal; the question was how far he would go in criticizing it. In what we regarded as a triumph, he came to Cadillac Square and advocated not the repeal but the *replacement* of the act. Our triumph pleased almost no one.

This willingness to yield, however reluctantly, to persuasion leads Martin (as it much earlier led McGowan) to argue that Stevenson's conservatism was also a pose. He was a liberal, but he did not wish to seem, even to himself, a man of catatonic response. He had to be persuaded, but he could be persuaded. After the 1952 campaign a series of seminars were held, with Adlai as the student, in Thomas Finletter's apartment in New York. These modernized his views even on economics.

I'm less sure than Martin as to the extent of this persuasion. I continue to think that the conservative vein was rather strong and that it was disguised by Stevenson's talent for insisting on his case with skill, charm and great good humor. Martin tells of Stevenson's return to Springfield after his first swing through the West in 1952. At a meeting that evening he warned those of us there that he was being pressed too far to the left by his staff; the time had come to get back — his words — squarely to the middle of the road. Incredibly, for nothing else is omitted, Martin fails to tell the rest of the story. David Bell, a liberal Truman aide, the Director of the Budget under Kennedy and now a high official of the Ford Foundation, was acting as chief of Stevenson's research and speech-writing staff. He protested, "You had a great response to your speech in Los Angeles, Governor."

Stevenson surveyed Bell sternly for a moment. (The speech had been to a large political rally.) "That crowd, David, would have applauded if I had advocated pissing on the floor." Then sensing that he might have bruised the man's feelings, he quickly added, "But that's something we should be for. Let's get it into future speeches."

One part of Stevenson's personality was always rightly read. He was a man of kindness, sensitivity and fun. And this, in turn, disposes of another misconception. Stevenson's marriage failed; he tried twice for the presidency and failed; he ended his life serving at the United Nations under a Secretary of State who was infinitely his inferior in stature and wisdom. It has been said that his was a disappointing life — indeed, a tragedy. That is far from the truth. Adlai Stevenson enjoyed both the campaigns. His personal life was rich in friendship. There was never a day without mirth — when something ridiculous didn't catch his eye and give him joy. There was no Stevenson tragedy. If there is to be sorrow, it must be for the country that didn't use him more wisely.

IV

...the Dubious Arts

24

Last Word on the Hiss Case?

FOR ANYONE born after 1940, the Alger Hiss case must be a major puzzle. There is the question of how he could have been considered guilty. Both the prosecution and the associated persecution were led by Richard Nixon. With such an enemy it's hard to suppose that Hiss could have done much that was wrong.

And there is even deeper wonder about why so much was made of the case, something contemporaries of Hiss and Chambers must themselves ask. Alger Hiss was a middle-level figure in the New Deal pantheon. He did not remotely rank with Haldeman, Ehrlichman, Mitchell, Kleindienst or Helms, the truly powerful malefactors in the later Nixon era, and perhaps not much above Chuck Colson. In the Roosevelt years a sharp distinction was made between routine functionaries and those called the Real New Dealers. At least by the war years Hiss was no longer a Real New Dealer. By many of his contemporaries he was thought polite, diligent but cautious, and somewhat of a stuffed shirt.

There was proof in his employment. When first hit by the House Un-American Activities Committee, he had just become president of the Carnegie Endowment for International Peace. The practical services of this organization to peace over the years have been one of the best-kept secrets of American life. The chairman of its board of trustees was then John Foster Dulles, and no one will now believe this meant a major commitment to

reducing great power tensions, avoiding dangerous brinkman-
ship or otherwise promoting international comity and tranquillity. The Endowment has, for long, in fact, been a highly reputable refuge for admirably respectable people lacking gainful
employment in the field of foreign policy. (There is never any
noticeable leadership, let alone any passion, for limiting arms
sales, the SALTs or countering the propaganda of the compulsive warriors.) Hiss had been at Yalta in a routine capacity, at
San Francisco in a more responsible administrative role, and he
had a moment of singular glory when he was pictured carrying
the United Nations Charter back to Washington or wherever.
But he was principally important because of his friends, a point
to which I will return.

If Alger Hiss was an accidental choice for historical eminence,
Jay Vivian (called by himself Whittaker) Chambers was an incredible one. In an age when Allen Dulles could accomplish
the Gary Powers U-2 disaster and the Bay of Pigs fiasco within
a year of one another, and E. Howard Hunt could, in one lifetime, mastermind both the Bay of Pigs and the Watergate operation (as well as the work on Dr. Fielding's office), we have
learned not to expect much of secret operatives.[1] But even by
such relaxed standards the selection of Whittaker Chambers by
the Soviets for that kind of work was luminously insane. His
past relation to the American Communist Party (where he had
shown sympathy for the Lovestone heresy) was one of demon-
strated unreliability. He was dangerously romantic and even
more dangerously given to self-glorification. After he went underground, he surfaced at frequent intervals to let his non-
Communist friends, most of them of compulsive literary habit,
know of his exciting way of life. And he was otherwise incapable
of accepting discipline.

I knew Alger Hiss slightly in the late war and postwar years.
I don't remember meeting Chambers, but I heard much of him,

[1] See a later essay, "The Global Strategic Mind."

for he was an editor at *Time* when I was, more briefly, one at *Fortune*. After Harry Luce, Whittaker Chambers was one of the major subjects of conversation in the community. People spoke of his literary ability and learning with awe, and his politics, then far to the right and paranoiacally anti-Communist, with dismay. He was often described as a little crazy.

Alger Hiss was charged not with espionage nor with being a Communist but with perjury for denying such matters. In the second of two major trials he was found guilty. While not everyone will ever be satisfied, Allen Weinstein, in his long-awaited book,[2] has gone as far as any historian could to establish the formal validity of the verdict. He has pursued witnesses to Hungary and Israel, information to estates, libraries and lawyers and into the files of the FBI. His treatment of the resulting material strikes one as both judicious and properly skeptical; he writes of it with clarity and restraint.

For many, the best Hiss defense was the suggestion that Chambers and the FBI had forged the typed documents by which he was convicted. This was accomplished, the defense suggested, by getting access to the Hiss typewriter or, when this seemed too improbable, by building another machine with the same idiosyncracies of type. Weinstein tells that a major effort was indeed mounted by Hiss's lawyers to duplicate the typewriter to show that it could be done. The effort failed.

For what it may be worth, I years ago concluded (as, I expect, did numerous other liberals) that to believe Hiss truthful was to strain the liberal faith too far. Only a conclusion that he had lied under oath would fit the vast and incredibly diverse and interlocking array of evidence. Weinstein has added greatly to the facts, and they force one to the same conclusion.

He has also caught without emphasis or false drama the mood of the time, of the trials that were so tensely watched, and also

2 *Perjury: The Hiss-Chambers Case* (New York: Knopf, 1978).

of the hyperbole and the red-baiting by the Hiss defense. Lloyd Paul Stryker, who led it, said of his client, "Though I would go into the valley of the shadow of death I will fear no evil, because there is no blot or blemish on him." And of Chambers he said that he was for "twelve long years . . . a voluntary conspirator against the land that I love and you love. He got his bread from the band of criminals with whom he confederated and conspired."

Professor Weinstein's contribution, then, is major, and I would say definitive. When one has lived with people and events and then later reads accounts of them, one almost invariably notices slips in the handling of initials, names, geography and minor happenings. It's a measure of Weinstein's competence that those who were around at the time will find few such errors in his book. There has been complaint that he misreported conversations with some of his sources. Perhaps he did, although people can alter their memory as to what they said. But, in any case, this complaint does not speak to the substance of his case.

So he has resolved one of the problems of the early middle-aged and young. To have Nixon as an accuser does not make one whole. Because it was not part of his task, Weinstein does not answer the other question: why did the case cause such an enormous fuss?

Communists were not remarkable in the nineteen-thirties. To be immune to doubts as to the excellence and success of capitalism in that dismal decade was to be unusually insensitive to the world around. For those who thought they responded to an uncluttered, unfearful, forthright mind, Communism was the obvious answer. Yet their numbers were always insignificant. By the time of the Hiss trials, it required a certain susceptibility to paranoia to believe that Communism was a serious force in the United States, its position in a few unions possibly apart.

Espionage would seem a more serious matter. But dozens of spies have since been arrested without achieving any similar distinction. Klaus Fuchs, the physicist and atom spy, was obviously a far more serious operator than Hiss, and, by comparison, he is now nearly forgotten. And it's a genuinely important as well as a much neglected point that Hiss spied at a time when there were almost no secrets. Few things can be so bad for espionage as a government in which everything is known. Much of what he passed on to Chambers for the Russians could have been had for the asking; more of it could have been had by listening to any one of the evening discussions that were compulsive in those days with those who were saving the country. I knew Washington well in the thirties and from within the government, and I can't remember ever having seen a paper stamped secret or hearing of materials so described.

In 1941, with the war at a critical point in Europe and our own participation imminent, my wife and I were visited in Washington by an Oxford-trained German friend named Jochem Carton, in later years the respected head of an important Montreal shipping company. He went one day to eat and drink with Rhodes Scholar friends and on the way back to our house felt urgent need for a men's room. He turned from Pennsylvania Avenue into the State, War and Navy Departments (now the Executive Office Building) and, inside, seeing an unlocked slatted door — a singular feature of the interior architecture of that aristocratic pile at the time — took it to be a toilet. It was, instead, an office, empty and with papers strewn over the desk. He tried the next door with the same result. Eventually he came to a door, this of solid wood, which did protect a toilet. That night he held forth on his experience with some vehemence; though an anti-Nazi who had recently become a British subject, he observed with emphasis that, with war in the offing, he was also about to become an ex-enemy alien. Such casualness, he thought, was incredible.

The wife of a onetime Soviet operative in the United States said of Washington during that period that one could have worn a placard proclaiming oneself a spy and not aroused the police.

What Hiss and Chambers were up to, Josephine Herbst, the novelist, thought, was "busy work," invented to engage the energies of the New Deal converts to Communism in a seemingly useful way. In fact, until well along in the thirties, the Russians in charge of covert operations in the United States did not think that Hiss or others similarly situated had any secrets that were worth the trouble. So the question remains: why the fuss?

It was partly, as many have said, that Hiss was a way of flogging F.D.R., the New Deal and the liberal Democrats. Numerous right-wingers had believed all of them subversive; the exposure of Hiss as a Communist and as a spy usefully affirmed the point. His exposure and conviction were also seen as a way of forcing a harder line on domestic radicals and toward Russia. But F.D.R. wasn't really in the line of fire. Nor were the New Dealers, many of whom were still in office. Nor, in any real sense, was the Democratic Party. At the most, the trials only stimulated the anti-Communist hysteria and witch hunts.

A further and better explanation concerns my fellow liberals and their anxieties. Allen Weinstein believes that Alger Hiss had divided his life into two compartments: on the one hand, he was the activist radical committed to world revolution as then perceived; on the other, he was the disciplined, motivated careerist, eager to be socially acceptable and a figure in what was soon to be called the Establishment. (Whittaker Chambers, it could be imagined, had a similar double existence: he was, at once, the romantic underground revolutionary and the learned, highly independent, compulsively articulate intellectual.) For what it may be worth, I have always imagined that there was an easier explanation of Alger Hiss: he was strongly subject to the fashions of his time, and, when disaster struck, he was in transition from the fashionable radicalism of the thirties to the prestigious foreign-policy establishment views of the postwar years, the Marshall Plan and the Cold War.

In either case, when Hiss came before the HUAC and to trial, he felt it necessary to deny his past. And the war and the post-war tensions with Stalin had given that past a seemingly new and greatly ominous dimension. Communism was now categorically wicked. Secrets, as the result of the war and the bomb, had become secrets indeed, and their security was now a ruling passion in Washington. Espionage had come in from the cold and was now called treason.

In the post-Watergate years, numerous politicians and corporation executives complained that once commonplace actions were now being judged by the post-Watergate morality. It was a defense for which Alger Hiss could have wished. And had he used it, it is interesting to reflect, he would have escaped. If, when he was first before Committee Chairman J. Parnell Thomas (soon himself to go to jail), Richard Nixon (eventually to be rescued by pardon) and the other exponents of the new Cold War morality, he had acknowledged that he knew Chambers and told of his strong Communist sympathies in the dark days of the thirties, of his fear of Hitler and fascism, of his consequent if deeply questionable help to the Soviets in pursuit of these beliefs, there would have been a major four-day furor. But when it subsided, he would probably still have been resting innocuously at the Carnegie Endowment.

But, instead, he agreed that the behavior of which he was accused was terrible, even unforgivable. This Lloyd Paul Stryker loudly proclaimed. Thus he denied his past. Since anyone, and especially a public official, leaves large tracks through life, such a denial can only be sustained by further improvisation. And this must be sustained by yet further lies. The task is of exponential difficulty, and the end, if pursued, is inevitable. Both Alger Hiss and Richard Nixon, his self-proclaimed nemesis, succumbed to the ruthless geometry of covering up the cover-up of a cover-up.

It was his condemnation of the actions of which he was accused and the denial that have given the Hiss case its impact. There weren't many liberals in the late forties who had been

Communists, and in the absence of secrets there were few Communists who would have been useful sources of information for the Soviets. But most liberals had *something* in their past — friends, suspect organizations, support for Loyalist Spain, sympathy for the Soviet experiment or a strong commitment to the wartime alliance with Russia and the hope that it would last. When Alger Hiss agreed that any such association was treasonable and otherwise totally reprehensible, he spoke for their past life as well. A whole generation of liberals was made to feel vulnerable. What most had to admit or deny, in the light of the new morality, was not very much. But, unlike Hiss, they would not have Justices Felix Frankfurter and Stanley Reed, Dean Acheson, William Marbury (for many years of the Harvard Corporation), Stryker and a half-dozen other great and famous lawyers as supporters, counsel or friends. If one so supported was threatened, who, however less his aberration, could be safe? And as the loyalty probes and investigations got under way and the blacklists proliferated, there was ample evidence that quite a few weren't safe. Additionally, by this time, a further few had, like Hiss, asserted, often with some indignation, that they, of all people, had never had a remotely pro-Communist thought, had never done anything even vaguely inimical to the sacred principles of free enterprise. They had also accepted a standard of past behavior by which they could themselves be impeached.

Here was the source of the interest, more precisely the anxiety, that made the Hiss case endure. It lived because thousands of intellectuals, reflecting on their past sympathies, actions or organizations, found themselves personally involved with it. And quite a few, by their disavowals, had set themselves up for similar assault, similarly provided themselves with glass jaws and, in consequence, were spending uncomfortable days praying, in most cases successfully, that they wouldn't be hit.

25

Bernard Cornfeld: Benefactor

PEOPLE ONCE BELIEVED that, given the lessons of the Great Crash and the Great Depression and the ministrations of the SEC, the days of truly fine financial levitation — on the scale of Ivar Kreuger as distinct from Bobby Baker — were over. Dull morality had set in. I had half thought so myself. It isn't so. Fools and their money can still be parted on as magnificent a scale as ever before. And with a certain artistry, as the story of Bernie Cornfeld proves.

Investors Overseas Services, IOS for short, was his command ship for a convoy of mutual funds, firms to manage mutual funds, firms to sell mutual funds, banks and insurance companies (including some shadowy financial figments that were so designated) and other more ethereal entities, all the brainwork of a Harvard-trained lawyer named Edward Cowett. Old Harvard men can take him as proof that we haven't been turning out only radicals; we still produce a very imaginative class of free enterpriser. The IOS mutual funds, which included the Fund of Funds which invested in other funds (although the other funds turned out, incestuously, to be mostly IOS funds), were sold by a sales staff that was orchestrated by another big operator named Allen Cantor. Statesmen of no slight distinction — James Roosevelt, once an excellent congressman; Sir Eric Wyndham White, a distinguished international civil servant and the longtime head of GATT; Erich Mende, once Vice

Chancellor of the German Federal Republic — graced the board, to their intense subsequent regret. It is said in their defense, and I think with truth, that they had no very accurate view of what was going on upstairs. Conducting was Bernard Cornfeld himself. The birth, rise and fall of IOS extended almost exactly over the decade of the sixties. Billions of dollars were extracted from the aspiring citizens of a dozen or more countries, with particular emphasis on Germany. (IOS resolved, in a manner of speaking, the old question of whether the Germans should pay World War II reparations.) One's sorrow for those who lost is tempered by the thought that so many suffered for their own cupidity.

For the rise of IOS was a testament to the continuing power of greed when reinforced by massive gullibility and a certain amount of spare cash. The warning signs were ample. It is elementary that no one should be trusted with money who does not convey, on all public occasions, an aspect of extreme solemnity. In behavior and general appearance, Bernie Cornfeld inspired somewhat less confidence than John Lennon, although more than Abbie Hoffman. Edward Cowett brought to the presidency of IOS a background of proven financial misadventure that would have seemed alarming to the late Samuel Insull. IOS was not allowed to do business in the United States, and eventually the SEC went so far as to extract from it a promise not to sell securities to any American citizen anywhere in the world. This far-reaching act of financial paternalism was well known. In 1966, Brazil, usually a fiscally tolerant jurisdiction, took numerous IOS salesmen off to the slammer for questioning. This completed, IOS was kicked out of the country. The organization was recurrently in trouble with the Swiss, who are also tolerant, and eventually it had to move its office operations from Geneva to a dismal village just across the border in France. Among other things, the Swiss took exception to an effort by IOS to pass a good part of its office staff off as students at the University of Geneva in order to get them work permits.

Yet IOS salesmen promised to make people rich, and people believed it and they bought and bought and bought. It was characteristic of the sixties, no doubt, that IOS professed some very elevated motives, such as bringing capitalism (the capital, not the exploitation) to the masses. And while it served extensively as a device for squirreling money out of the less developed countries and into some of the least promising of American enterprises, it pictured itself as a powerful instrument of enlightened economic development. People who did not lose money will be relieved to know that capitalism still functions according to the old rules. If Cornfeld and Cowett had, indeed, been the instruments of spiritual and moral regeneration, a great many other assumptions about capitalism, capitalists and their motivation would require revision, and Gideon Bibles would be enclosed with annual statements.

In fact, IOS was a gigantic sales organization operating at very high pressure. And that was about all. Its salesmen not only sold the funds (or more precisely, undertakings to buy into the funds over a period of years), but they also sold other salesmen on the idea of selling funds. So, in time, a salesman graduated from collecting commissions on his own sales to collecting on the sales of those whom he had sold on selling. And then with more expansion he would collect on the sales of those who were collecting on the sales of those who had sold yet others on selling. (By the end, this pyramid was around six stories high in Germany.) The salesmen and the various hierarchs who were called sales managers were also rewarded with the right to buy shares in IOS. Many, believing devoutly in their own sales spiel, borrowed money to do so. When the shares of IOS became more or less worthless after its crash, they suffered a fate somewhat worse than that of their victims. Their shares became wholly worthless. It is a fixed feature of all such disasters that many of those involved expropriate themselves.

The surviving worth of the underlying funds was not excessive, for the large cost of selling was deducted before the cash was invested. So were other charges. With the passage of time also, IOS tried to improve the value of its funds by investing heavily for performance, and it hired and rewarded investment managers in accordance with their ability to get capital gains. This is approximately the same as saying that it rewarded them for finding insane investments, for that is where the chance of capital appreciation (at least for as long as a particular stock was attracting the attention of other gunslingers) seemed to be. No similar magic could be expected from AT&T or Shell. In the latter days of IOS, investment even took the form of heavy loans to its own executives. Sales had to increase to cover the steadily increasing costs of the sales operation and also the high living costs of the IOS management. The stock market slump in 1969 put the handwriting on the wall, and the crash brought down the wall. Robert Vesco, as the next essay tells, then came in, and the real looting began.

It would be fun to go on telling the story. One John King of Denver, Colorado, made a fine appearance in the play, and the collaboration of King and Cowett in writing up the value of exploration rights on a vast stretch of Arctic oil lands (and ice) was a little masterpiece of its kind. A small acreage was sold at a very high price; this was then taken as justification for the revaluation of exploration rights over a huge water- and landscape. IOS then collected a performance fee on the increase. It deserved it, for the original sale was a sweetheart deal arranged between insiders to inflate the value of this indubitably and literally frozen asset. Toward the end, IOS was an enormous buyer of its own securities. That, for those who sold, was a very charitable action. It is also the common and costly act of desperation of promoters for whom the end is in sight but no more visible to them than to their victims.

The history of Cornfeld and the IOS has been written with competence and restraint by three talented British reporters, Charles Raw, Godfrey Hodgson and Bruce Page.[1] They bring Bernie Cornfeld's girls, castles and planes into the story mostly as these contributed to expense. They refrain from throwing up their hands or gasping with indignation. A great deal of time and patience is given to unraveling the more intricate of Cowett's deals. I have only one small personal quibble.

In the early summer of 1967, Cornfeld provided the funds for Pacem in Terris II. (It now seems that he didn't pay but instead put the bite on his brokers.) This was a gathering in Geneva of scholars, journalists, divines and professional men and women of goodwill to urge the case for peace in general and in Vietnam in particular. Justice William O. Douglas, Martin Luther King, Jr., Senator Joe Clark, were among those who attended. The authors think those present, of whom I was one, were a bit innocent and must have been embarrassed when they knew more of the auspices.

I don't think that's so. Although they are not breathlessly interesting, such convocations are useful, for they remind people, the press and the politicians that peace also has a constituency. There is more than NATO, Lockheed and Rockwell International. However, I specifically raised the question of the auspices with Robert Maynard Hutchins, who was running things. He replied that those who beg on behalf of good causes can't look a gift horse in the mouth. The metaphor aside, I think he was right. I did take the precaution of paying my own way, but that was more out of prudence than principle. So we were not all uninformed.

One reason I tended to be suspicious was that I had previously written about the 1929 crash, and IOS looked impressively like a replay of the work of Goldman Sachs and the other public benefactors of the period. Also, I go frequently to Switzerland to write (and to Geneva where I am a largely honorary professor), and I had heard the Swiss bankers on IOS. They ranged from denunciatory to malevolent. Most un-Swiss. And finally Jimmy Roosevelt had once journeyed to Cambridge to

1 *Do You Sincerely Want To Be Rich?* (New York: Viking Press, 1971).

sound me out on joining the Board of IOS. I handle financial matters with care, better on the whole than does the average professional investment man. But my effect on the nerves of people with money to invest is often adverse. Any financial concern proposing to make me a director, and for window-dressing at that, was surely suspect.

26

Robert Vesco: Swindler

No SPECIES of literature has given me so much pleasure over the years as that celebrating high-level fraud. I never miss a book on the subject; the occupationally most obtuse publishers have learned this and send me proofs of work that is often barely literate. I read it all and praise it all, for I enjoy all of it.

My pleasure comes partly from working out how the fraud is done. This is a simple, undemanding form of puzzle, for there are only a very few basic forms of financial graft, on which, in turn, there are a variety of engaging themes. By far the largest family are the pyramids which can exist in time, space or both. All collect from many, pay something out as bait and concentrate the rest on the swindler. All pyramids involve nice judgment as to when to get out, something that is beyond the power or discrimination of almost every crook. So, with rare exceptions, swindlers are open themselves to the deception they are employing. They end up swindling not only the public but themselves.[1]

The pyramid in time is the Ponzi operation.[2] Earlier investors are paid off with the money invested by later ones. The later take must grow exponentially if it is to cover the increasing pay-off required by the earlier gulls. Thus the pyramid. The trick is to get to Brazil before the pay-off is too big in re-

[1] As indicated in the preceding essay.
[2] See p. 325.

lation to what is retained by the swindler. The most recent
example of the Ponzi pyramid was Home-Stake, an operation in
nearly nonexistent oil; its variation was in using men of extreme financial sophistication, up to and including high officers
of Citibank, as unknowing shills for people, mostly from show
business, who laid claim to no financial sense whatever, only
to more money than they could use. The first in were paid
off by those who came later until not enough came. There
was a delightful aftermath to the Home-Stake story. A *Fortune*
writer, reporting on the enterprise, denounced a new and exceptionally depraved tendency in public regulation; that was
for regulatory agencies to concentrate on protecting only the
poor. The Home-Stake saga, he said, with indignation, showed
that the SEC "was interested in the small investor, not the
rich." SEC officials denied there was "any such discrimination."
I take up Home-Stake in the next essay.[3]

The franchise operation is a pyramid in space. A salesman
recruits other salesmen and gets an override — a share in what
they receive as down payments. And then the salesmen so recruited recruit yet other salesmen and take a similar cut before
passing the residue back to the individual who recruited them.
As long as the operation is expanding, the salesmen are greatly
inspired to sell other salesmen on selling. And revenues pile
up back on the originator. The trick is to go public and unload
the stock in the ultimate company while the expansion is still
going on. For when recruitment and sales fall off, revenues are
absorbed along the way. Not much or nothing gets on up to the
ultimate company.

The holding company or investment trust pyramids are a
pyramid in both time and space. Debentures or preferred stock
are issued, along with common stock, to buy into yet other
firms, which buy into yet others. In the late nineteen-twenties,
before the law intruded, this was repeated through half a

---

[3] "Should Stealing from the Rich Be Punished?"

dozen or more layers. As long as the earnings or capital value of the ultimate company go up, these accrue not to the pre-ferred-stock- or debenture-holders, whose returns are fixed; they concentrate upstream on the holder of the ultimate common stock. This is leverage, and in every boom the wonders of lev-erage are rediscovered. The thing here, of course, is to get out while earnings and values are going up, for when they go down, all values and more are absorbed as collateral by the senior securities along the way. Nothing is left for the common stock and certainly none for the ultimate common stock.

Again almost no one involved in this kind of swindle gets out in time. All see in the rising stock and the way the pyra-miding is accomplished a reflection of their own intelligence in stock selection and corporate design. Financial genius, it cannot be said too often (although I have tried), is a rising market.

As I've described in the preceding essay, Bernard Cornfeld's achievement in IOS, his girls with the tired eyes apart, was a multiple pyramid, an admittedly complex metaphor. He com-bined the pyramiding in sales operations earlier noted with the financial pyramiding just described. IOS rewarded itself according to the performance of its funds. The principal fund, the Fund of Funds, got performance by investing in other mutual funds that featured performance, which invested, then, in firms that featured rising values. This greatly rewarded IOS as long as there was performance. Thereafter there was nothing.

There are swindles other than the pyramids, but they are not very interesting — peddling of worthless securities, em-bezzlement and corporate looting. Equity Funding, the great California insurance swindle, although at first glance it seemed new, was really a Ponzi operation. The insurance companies were being gulled into buying insurance contracts on non-existent souls out of Gogol. They were paid off from the initial proceeds of the sales of yet more policies on equally nonexistent souls. The fraud was partly concealed by computer operations which the state regulatory agencies did not understand, by a good deal of creative accounting and by hiring people to fake the contracts who mostly didn't know what they were doing.

Along with the simple pleasure of identifying and classifying the type of crime, there are the very distinguished people that one encounters in this kind of work. I first became aware of this reward some twenty years ago while searching out the details of the great Goldman Sachs expropriation of the late twenties; the most prominent name associated with this rape of the innocent, an association I have celebrated elsewhere,[4] was none other than the God-fearing, or anyhow God-invoking, John Foster Dulles. By the time of my study, nonvirtue had reaped its frequent reward. Dulles was Secretary of State.

Other and equally gamy operations of the late twenties enlisted some of the more saintly economics professors of the time. In later years the New Dealers, with some exceptions, steered clear of such involvement, perhaps partly because of lack of opportunity. But in the boom of the sixties there was a dismaying migration from the New Frontier into the more egregious of the offshore funds. Many of those so moving were believed to be inspired by a strong sense of public duty. Having done good in Washington, they wanted to help the poor and abused of the world become rich through sage, imaginative use of their savings.

In the age of Nixon there was a new crop of swindles and a new crop of names, although the Nixon men, to their credit, had unmixed motives; they were not under the impression that they were doing the poor any good. The most depraved faces from this period were those surrounding Robert Vesco.

The story of Vesco has been told by Robert A. Hutchison, a careful man who writes well, with just the right note of contempt for all concerned.[5] Vesco moved in on Cornfeld when IOS seemed to be on the verge of collapse in 1970, and he pro-

---

[4] In *The Age of Uncertainty* (Boston: Houghton Mifflin, 1977), pp. 208–209.
[5] *Vesco* (New York: Praeger Publishers, 1974).

ceeded to ensure that collapse. The financial pages of the news-papers, from incompetence and gullibility the nearly invariable allies of the crooks, described him in terms that he had largely invented for himself — a dynamic New Jersey financier and industrialist who, one would assume, was in daily touch with David Rockefeller. A single candid column would have alerted the SEC, maybe the IOS board of directors and possibly even the local police, for Vesco was, in fact, a ragged-ass operator whose assets were as exiguous as his commitment to truth. IOS, however, was his opportunity. By 1970, the investment pyramid and the sales pyramid had both collapsed and therewith its revenues and the value of its stock. Fixed charges continued. Those involved reacted not like accomplished criminals but like frightened juveniles. Vesco showed up and agreed to pour in new money in return for effective control. Bernard Cornfeld seems to have been one of the few who recognized that however bad he himself might be, Vesco would be a hundred times worse. Over his objections Vesco got control for virtually noth-ing, and certainly nothing of his own. The rest was inevitable.

For Vesco was not a pyramid artist but a simple, dull looter. Before he arrived, not too much had happened to the vast pools of money that had been assembled for investment by the funds that were mismanaged by IOS. Values had shrunk; a fair amount had been dissipated in lunatic efforts to achieve or simulate performance. Some of the wealth, as noted in the last essay, was from writing up the value of a vast acreage of Arctic oil land, terrain which suffered from, among other handicaps, not being land at all but ice or very cold water. The collapse took even less time than the write-up. Yet substantial assets remained. Vesco proceeded to get these out of their previous custody and reinvested in corporate shells, in which, not sur-prisingly, the primary ownership interest was with Vesco. His larceny was not especially deft or artistic. It was remarkable only for the sheer bulk of the assets so transferred and seques-tered. A really accomplished crook with any pride in his work would not, I think, have wished to associate with him.

Those who did get involved were low types, including John Mitchell, Maurice Stans and numerous Nixons. Vesco's plan was to loot two or three hundred million and fend off the SEC by buying up the Nixon family and the Republican Party. So he invested in both of the President's brothers, Edward and Donald, as well as the son of Donald Nixon, also called Donald. This last Donald, Vesco carried around with him as a combined talisman, hostage and souvenir. Then he put $250,000 into CREEP — the 1972 Committee to Re-Elect the President.

For a while it all seemed to work. Maurice Stans served him as impartial ombudsman for a citizen with a problem; so did John Mitchell and John Dean. As Attorney General, John Mitchell got Vesco sprung after the Swiss, far more perceptive than we, tossed him in the lonesome when one day, through an error in judgment, he showed up in Geneva. In the end these men found themselves in more trouble than Vesco, for he moved on to Costa Rica and they were forced to stay.

Part of my delight in this whole subject matter is the occasional encounter with a truly splendid rascal — someone in the tradition of Jim Fisk, Jay Gould and Ivar Kreuger. Vesco wasn't. But he had one quality that commands admiration. Most of those who kicked in the big money for Nixon and the Republicans, when required to explain, said it was because of their deep commitment to American democracy or their desire to save the country from George McGovern and socialism. That they should want anything for themselves — well, perish such acquisitive thoughts. Vesco, on the other hand, when he handed over the $250,000, made himself, as Richard Nixon often said before obscuring an issue, perfectly clear. He only wanted protection from the SEC. No nonsense about saving democracy or even free enterprise.

27

Should Stealing from the Rich
Be Punished?

ALTHOUGH I DON'T KNOW HIM personally, I have for several years been an admirer of Mr. Walter B. Wriston. Partly, of course, it is for the huge amount of money over which he presides as the head of Citicorp, operator of the nation's second largest bank. Partly it is for the impeccable information that he often conveys to the public. At a meeting in London not long ago he began by reminding his audience that "Much has been said about the future fate of all of us who live on this planet." (It does seem wise to omit reference to the excessive volume of writing on man's past fate and also to the highly unreliable literature on life on other planets.) He went on to say, also truthfully, that "Today, as always, there is good news and bad news." But no man is perfect. Instead of stopping while ahead, Mr. Wriston proceeded to assail critics of business for saying that it is "what 20th century mathematicians call 'a zero sum game,'" which is to say that a profit for one chap always means a loss for someone else. Perhaps only momentarily and because only two hundred-odd grand was involved, Mr. Wriston had forgotten that he personally had sat in on one of the greatest zero sum games of recent times, the great Home-Stake swindle, and that historians being what they are, he might be even better remembered for this than for his thoughts on good and bad news, although I hope this won't be so.

I want to get away from Mr. Wriston and on to the swindle, but I have to add that he showed in this whole episode not only that the head of the second largest bank in the country can have the common touch, be as competently gulled as anyone else, but also that, for a charitably minded man, the zero sum game can, like poker, have a floating version. Mr. Wriston gave part of his participation in his excessively nonproductive, sometimes nonexistent, oil wells to the Fletcher School of Law and Diplomacy at Tufts University. According to university sources, the Tufts gift was valued for income tax purposes at around $38,000. The ultimate return to the university was not quite zero; Tufts realized $100. That episode and associated deductions may well confirm another of Mr. Wriston's truths, one that was quoted in the *New York Times* on August 2, 1977. "People with equal incomes," he declared, "pay unequal taxes."

The details of the ride on which the great banker was taken (and which he so thoughtfully shared with universities and the sick) are from David McClintick's *Stealing from the Rich: The Home-Stake Oil Swindle*.[1] McClintick has done well; the only conceivable fault with his book lies in the title and possibly in its approach to punishment, to which I will come later. The title, as the author admirably establishes, should have been *How the Rich Swindled Each Other and Themselves*. More passive victims were the United States Treasury, which got hit by a large number of dubious tax deductions, and numerous charitable and educational institutions (including the Library of Congress), which were given highly valued but worthless investments, mostly, it would appear, after their generally ethereal nature had been perceived by the various angels of mercy who bestowed them.[2]

*

[1] (New York: M. Evans and Co., 1977).

[2] In 1971 and 1972, the Library of Congress received gifts of Home-Stake participations valued for charitable purposes at $485,750. By the end of 1974, these should have yielded the Library between $59,500 and $119,000. Actual returns were $1660.87.

The enterprise that Mr. McClintick so elegantly describes got under way in the mid-nineteen-fifties, the work of a lawyer-entrepreneur named Robert Simons Trippet of Tulsa, Oklahoma. Mr. Trippet combined a major instinct for piracy with, most exceptionally, a careful, rather conservative intelligence. It was his belief or discovery that the financially highly solvent get most of their investment information from each other, do not investigate, do not or cannot read, are otherwise deeply retarded on financial matters and would rather lose a large sum of money than pay a smaller amount to the government. On all points he was proven right. In pursuit of these principles he learned that to swindle the rich, you don't need to do very much, you only need to do it very well. You must, in particular, never forget that it is style, not substance, that counts.

As I've argued in an earlier essay, there are no new forms of financial fraud; in the last several hundred years there have been only small variations on a few classic designs. Trippet's fraud is that associated with Charles Ponzi, an energetic Boston operative who, in the early twenties, took hundreds of affluent fellow Bostonians to the cleaners. But the same design was present in the operations by which John Law bilked an even larger number of Parisians between 1716 and 1720, and there are many indications of its far earlier use. It consists, simply, of paying off earlier investors with the proceeds of the sales to the later ones. The earlier ones, because they are getting a handsome return, bring in the later ones, sometimes with an incontrollable rush. Trippet's variation made use of the tax laws. He sold participation rights in the drilling of sometimes hypothetical oil wells which, had they all been real and had the costs actually been incurred, would have allowed the investor a legal, although quite possibly unjustified, tax deduction for the whole amount in the year of the expenditure. Trippet carefully and selectively rewarded the earlier investors from the later investment flow and, on occasion, returned the original investment. His special concern was for those who were most likely to advertise either their good fortune or their loss.

He here raised to the level of high art another classic financial technique, namely the conscientious greasing of the squeaking wheel. He also drilled some wells so there was something that any abnormally curious investor could see, although at the Santa Maria Field in California some were shown imaginatively painted irrigation pipes which they were told involved oil. Most important of all, Mr. Trippet knew the value of name-dropping. He pyramided both money and names.

His method, which may have worked better than even he could have foreseen, was to get two or three gulls with a reputation for great financial wisdom in each of a half dozen intellectually and socially incestuous banking, business, social or professional circles — the close-in convocations in which people regularly convey highly unreliable information to each other, often in the strictest confidence. While these gulls were telling each other that they were on to a good thing, Mr. Trippet and his small coterie would help the process by dropping in the names of their leading shills. This was a service that the Wriston name rendered when dropped in the New York financial community, although there was much sincere help from George S. Moore, Wriston's predecessor at Citibank, Hoyt Ammidon, head of the United States Trust Company, Thomas S. Gates, former Chairman of Morgan Guaranty, and Reese H. Harris, Jr., executive vice president of Manufacturers Hanover Trust Company. Such names fall among bankers with an infinitely reassuring clang. Hoyt Ammidon, a pleasant man whom I know, was an especially poignant name for me, for not only did he get taken for $218,500, but his company has looked after my wife's and my small financial affairs for around a third of a century. This they have done with intelligence, integrity and a pleasant freedom from loss. Our only serious difference of opinion came a few years ago when they proposed that we interest ourselves in oil, cattle or other tax shelters. I rebuked them, not because I knew anything about Home-Stake or other shelters but because it occurred to me that if the risk weren't excessive or there weren't some other flaw, no one would be paying any taxes at all. It was also my deepest conviction that no New York banker could possibly know anything whatever about cattle and not much about oil.

What Moore and Wriston did as pilotfish for the bankers, Fred J. Borch, former Chairman of General Electric, did for industrialists, and he was terrific. When he swam in, so did a whole school of GE men. In an appendix to his book, Mr. McClintick lists the principal victims by industrial, legal or other occupational groupings. There were so many GE vice presidents and vice chairmen, along with a treasurer and a couple of chairmen — thirty-five in all — that they make up a special class by themselves. Other heavy and light industry is well, though less munificently, represented. Present also are two former Secretaries of Defense.

A third community was the great New York and Washington lawyers — members of the big firms that specialize in keeping the big corporations solvent as well as honest. The Washington lawyers seem to have suffered from the guidance of an old friend of mine and fellow economist, Dr. Redvers Opie, a onetime member of the Harvard faculty and later a fellow of Magdalen, one of the most cherished of Oxford colleges. Redvers spread word of the wonders of Home-Stake around Washington and received a modest commission whenever, in consequence, some affluent legal sucker hit the bait. He seems, by the best possible evidence, to have been wholly innocent of any larcenous intent. He, in fact, put his own money into Home-Stake and appears on Mr. McClintick's list of losers.

There was also a Los Angeles community of business and financial fleecees, and there and in New York was the most alluring, romantic and enchanting group of all, the great entertainers and artists. Among these lambs the slaughter was simply appalling. Jack Benny, Candice Bergen, Faye Dunaway, Bob Dylan, Mia Farrow, Liza Minnelli, Mike Nichols — the list goes on and on, and for many the amount was up in the Ammidon range. For any first-rank performer to be off this list is to arouse real suspicion; unless terrific alimony is the explana-

tion, the individual was working way below rates and faking
his or her salary to the press. Barbra Streisand's name appears
so impressively in the book that when the author mentions
Santa Barbara, the printer spells it Barbra too.

Of the established forms of graft, none has been so common
in Hollywood for so long as the looting of motion picture in-
nocents by their business managers or sometimes their lawyers
and brokers. I've known many Hollywood folk over the years,
partly out of interest in what they do, partly in our common
pursuit of Democratic politics and contributions. They have a
tendency to pour out their personal financial history to an
economist; this, for almost all, includes memory of some massive exploitation by a hitherto deeply trusted, deeply larcenous
business adviser, manager, agent or friend. Home-Stake was in
a great artistic tradition.

Three things protect the simple-minded investor from thiev-
ery under our system. These are the watchful eye of the independent auditor, the menace of the law and the courts and the
supervision of the Securities and Exchange Commission. Over
the nearly twenty years that he survived in fraud, Trippet was
only mildly bothered by these impediments. Auditors repeatedly uncovered or suspected thimblerigging and theft; if they
persisted in their suspicious way, Trippet simply got others of
more agreeable disposition, denser mind or greater need for
the pay. Once or twice his highly creative annual accounting
was issued without benefit of an auditor's attestation. No one
much noticed. As the years passed, he was sued by investors
who had come, however gradually, to realize that they were being scammed. The lawsuits attracted virtually no attention
outside of Tulsa. And, in a nice gesture to his friends and
neighbors and perhaps also because they were insufficiently innocent, he largely refrained from stealing at home. The SEC
was a risk; it requires, as all know, that future investors be advised through a prospectus of the risks they are running. Trippet, in his prospectuses, never got around to telling the whole

truth, but he did tell quite a bit. Thus the 1970 admonition, more explicit than those that preceded it, said:

THESE SECURITIES INVOLVE A HIGH DEGREE OF RISK AND THE EXISTENCE OF THE POSSIBILITY OF SUBSTANTIAL COMPENSATION TO THE OPERATOR [Home-Stake]. To date, none of the investors in any of the prior programs which Home-Stake and its subsidiary operators have offered to investors have recovered their entire investment.

Trippet's guess, however, was that a fool who really wanted to be parted from his money would not read or would not understand such warnings. He was right.

At long last, in 1973, in close step with the Watergate uncover-up, the Home-Stake fraud did unravel, and in the following year it came to pieces. The general press, over the years, had shown itself, as always, to be worthless for searching out financial hocus, even when this was massive and practically overt. The head of any great news-gathering organization who was promised a return of several hundred percent on his investment should immediately have alerted editors, reporters and researchers to find out what was going on. Such intervention with the editorial side by the corporate brass is never criticized. Alas. James R. Shepley, the President of Time, Inc., got such a promise of improbable wealth, ordered no investigation, but invested himself. It remained for the *Wall Street Journal,* with its subversive instinct for assembling dull facts, to reveal the true nature of the Trippet expression of the free enterprise system. That was a worthy act; it had the further great advantage of bringing McClintick into the play, for he is a *Journal* reporter and, in consequence, wrote this book.

Brought to the bar of justice, as it is often called, Trippet pleaded no contest to charges that he had performed a total of some ten felonious acts at the expense of the trusting. U.S. District Judge Allen E. Barrow, who presided over the trial, which featured a generally inept prosecution, distinguished himself for both his sympathy for the wrongdoers and the deeply tortured character of his prose — "... certain aspects of the so-called Home-Stake trial have become clear to me as the trial judge. Not the least of these is that the whole thing is not what was pictured by some Eastern newspapers and magazines in their colorful or sensational reporting, however you want to call it, as the century's greatest swindle. . . . I have seen evidence that there was a grandiose promotional scheme that went sour . . . These men admit having done wrong, having violated the trust that the individuals reposed in them. One must accept these confessions of fault and abide by the judgments upon themselves thereby pronounced." To this possibly bearable personal judgment, the judge added a maximum fine, ordered a contribution of $100,000 to a fund for any stray widows, orphans and children who had been rendered destitute by Trippet's operations (they had not been a rewarding target; they didn't have enough money and, in contrast with the New York bankers, may have been too prudent in handling what they had), but confined more sanguinary punishment to the jail time already served. That had been one night in the local cooler. Trippet wrote a note of warm thanks to the judge.

David McClintick is critical of Barrow. However, his awful prose apart, there is something to be said for him. The people of the United States were cheated by the high value given to valueless gifts and the consequent tax deductions. As a participant citizen and resulting victim, I want to see every effort at recovery and I hope the IRS will be held closely to account on this. But the judge was only marginally involved with this aspect of the case. He was more concerned with the original swindle, and it may be that people who don't treat their money with intelligence and respect, who have so much that they don't need to do so and who are willing to entrust it to any scoundrel as an alternative to paying part of it to the government are beyond the protection of the law. Had Trippet not taken them, it would have been someone else. Maybe Judge Barrow was right to go easy on the man; he merely managed to get in on the top floor first.

28

The Global Strategic Mind

In the summer of 1978, there was a small but interesting explosion in Washington over the effort by an old Harvard colleague of mine, Samuel P. Huntington, then on assignment to the White House, to get Senator Daniel Patrick Moynihan, also recently of Harvard, to make a public assault on a presidential decision involving trade with the Russians. I was sorry to read about this. I had repeatedly urged my Harvard colleagues, when they go to Washington, not to practice the kind of politics which is commonplace in Cambridge. Washington is not ready for it. But I was much more alarmed by an earlier communication from Sam Huntington in which he defended a briefing on strategic balance that he had given to the Chinese. It involved, he said, no secrets; it was one of many such briefings he had been giving. This was an indication that global strategic thought, sometimes called relentless strategic thought, was again rampant in Washington. There is nothing about which the country should be more concerned. President Carter can shrug off the odd professorial attack. But all recent experience shows that Presidents cannot survive the strategic mind. If life on this planet dissolves one day in an intense sheet of flame with great overpressure, the guidance to our demise will have been given by a relentless strategic mind, a particularly tough exponent of global balance.

*

The strategic mind is readily identified and, on the whole, rather simple as well as straightforward. It is drawn uncontrollably to any map of the world, and this it immediately divides into spheres of present or potential influence. The nature of the influence is never specified. Nor are the consequences of its exercise, if any, except as there may be vague references to essential raw materials, naval bases or the control of adjacent waters by local aircraft — threats that, in the end, do not materialize, perhaps because they only rarely have anything to do with modern military need or technology.

A hostile sphere of influence always requires a prompt and lethal reaction. That is partly because nothing else will be understood by our opponents and partly because it must be shown that we are capable of such a reaction. Those who question the need and ask consideration of the consequences show by their insistence on thought that they are indecisive. The man of relentless strategic mind substitutes bravery for thought and action for reflection.

Above all, it is an essential of strategic thinking that it ignore experience. That is an intensely practical matter. Once the strategic mind starts reflecting on experience, as I will presently urge, it is down the drain. Experience is the one thing by which it cannot, under any circumstances, be guided.

The strategic mind manifests itself among the scholarly classroom warriors at Harvard; rather harmlessly in onetime summer soldiers and diplomats in the New York corporations and law firms who meet nostalgically at the Council on Foreign Relations to recall their days of glory; in the Pentagon, the CIA and, though more fugitively, in the State Department; in the National Security Council, alas; and it flourishes among the housecarls of Senator Henry Jackson. It is disastrously represented in other countries, the disaster being, as usual, for the governments it guides. But let me recur to the experience that, professionally, it is required to ignore.

*

The most vital recent experience was, of course, in Vietnam. For a little while after the war, it must be said, the strategists were a trifle subdued. Hundreds of thousands of ethnically diverse lives and many billions of dollars had been spent to keep Saigon out of the Chinese sphere of influence. There had also been concern about the Russians, but the Chinese were the nearer and more dominant force. It had been said that in the absence of our effort, the dominoes would fall. Chinese Communist influence would extend across Thailand to Malaysia, Singapore and beyond. Now, however, the Vietnamese and the Chinese have fallen out. Chinese assistance to Hanoi has been stopped. Chinese long resident in Saigon are being expelled, and there has even been fighting on the border. As for the Russians, they face a bitterly hostile China, which increasingly we befriend. The men of strategic mind concede that there is a lesson from Vietnam: we must not allow it to weaken our will to intervene the next time.

These global strategists also complain that Vietnam still dominates our thoughts to an excessive degree, and in this I concur. There should be much more thought about Africa, for that has long been a positive addiction of the strategic mind. It is there, above all, that we need to reflect on past experience.

Thus, in the nineteen-sixties, there was a deep strategic concern over Tanganyika *cum* Tanzania, where the Chinese were building a railroad; there was even greater worry over Guinea, which had been all but written off to the Communists; and in the early days of the Kennedy administration there was a near frenzy over the Congo, now Zaire. The latter, we were all told, we simply could not afford to lose. In time, strategic intervention having been resisted or kept ineffective, the Tanganyikan railroad was forgotten; the Soviet ambassador got kicked out of Guinea, where we are now mining the bauxite; and Zaire lapsed into a system of eclectic, universal but not especially innovative corruption. As a footnote proving that the strategic mind is ever resourceful: In 1961, when as ambassador to India

I first called on Nehru, he expressed concern about one Clare Hayes Timberlake, then our ambassador in the Congo. He thought him to have an unduly sanguinary strategic mind. Timberlake had served previously in India where so great was his strategic commitment that he had endeavored to keep Nehru out of the hands of the Communists. Nehru had not thought it necessary.

All of this was overture. In the next years the strategic minds became alarmed over Algeria, a major bridgehead into Africa, where Ahmed Ben Bella was being advised and supplied by the Soviets. Then one day in 1965, Colonel Houari Boumedienne took Soviet tanks into Algiers and arrested Ben Bella, who, having spent the Algerian war in jail, was now put permanently back in confinement. A very restricted life. A member of the Soviet Embassy in Washington whom I asked about this development said, "At least they didn't use our advisers."

In these years the exponents of global balance also expressed great concern over Soviet influence on Kwame Nkrumah in Ghana. The Soviets were training his praetorian guard, were influencing his economic policy and were thus a threat, although this was not stressed, to our chocolate supply. In 1966, Nkrumah was thrown out; the worry had been to no purpose.

Concern then shifted to Egypt, where earlier John Foster Dulles, the modern prophet of all strategic minds, had paved the way for the Russians by trying to accommodate Gamal Abdel Nasser to our sphere of influence in the matter of the Aswan Dam. Soviet aircraft, anti-aircraftmen, advisers, some thought even pilots, had thereafter swarmed in. The strategic minds were at the point of paranoia; at a minimum we should encourage the Israelis to put an end to it all. Then the Russians were all sent home by the Egyptians.

Of late, in that general part of Africa, attention has been on Ethiopia. This was once a bastion of the free world; now Ethiopian military men whom we trained have come dangerously under Soviet influence as they fight against Somalia, once a

Soviet enclave but now itself a bastion of the free world. In one powerful strategic view, Somalia must not be defeated because Ethiopia, and thus Russia, would then dominate the oil routes into the Red Sea. This is where long-range guns come in, one gathers.

To the south and west in Angola, the strategists, earlier on, had translated their alarm into actual intervention. The former Portuguese colony was by way of becoming a Cuban and Soviet colony — to the strategic mind a quite plausible exchange. And it would dominate the supertanker routes to Europe — more of those long-range guns. However, our goal was not to keep the Cubans and Soviets out but to make that imperial effort as costly as possible and to prove that, after Vietnam, we were still capable of response, however insane. This story has been told in impressive and convincing detail by John Stockwell,[1] a former field-grade officer of the CIA. It should not be missed. Since strategic thought survives by ignoring experience, it has a highly professional interest in avoiding accounts such as this. By the same token, all who are alarmed about the tendency toward such strategic thinking should welcome Mr. Stockwell's book, which is both well-written and interesting.

The usual final word is now emerging on strategic thought in Angola. Evidently that country wasn't lost either. Its government and economy are being sustained all but exclusively by money from Gulf Oil. That sort of thing takes the edge off a Communist victory. And we are moving to establish friendly relations. One sees again why the global strategists must ignore experience. We have won only when we have resisted the impulse to intervene.

In the old witch-hunting days it was a routine precaution, on uncovering American overseas aberrations, to come up with some counterpart Soviet mischief or misbehavior. That is no

[1] *In Search of Enemies: A CIA Story* (New York: W. W. Norton, 1978).

longer essential, but there is no escaping the fact that the relentless strategic mind, pervasive and omnipotent, functions in the Soviet Union as the mirror image of American error. And then vice versa again.

Thus it was once central to Soviet strategic thinking that China was part of a wonderfully expansive sphere of Soviet influence. To this end technicians and money were dispatched in quantity. And our strategists saw eye to eye with the Russians. Dean Rusk called China a "Soviet Manchukuo," lacking in even the barest essentials of sovereignty. From this position he retreated to argue that any division between the Soviets and the Chinese was only over the tactics for destroying the free world. For many years James Angleton presided with what seems to have been clinically acute paranoia over counterintelligence activities. William Colby, former head of the CIA, in another book which, sometimes unwittingly, illuminates the workings of the strategic mind,[2] says that Angleton never accepted the Sino-Soviet split as genuine, "at least until a few years ago." It was, Angleton avowed, an artful Communist way of misleading the nonstrategic minds.

Returning to the Soviet strategists, were it professionally permissible for them to reflect on their experience, they would have to concede that Indonesia and Sukarno were an investment gone sour. And likewise, of course, Ben Bella. And similarly Nkrumah. And in an especially costly way also Egypt. And similarly Somalia. And now possibly Angola. Nor is this all. India under Indira Gandhi was held by the best American strategists to be moving dangerously into the Soviet orbit. The Soviet strategists must have been as pleased as ours were worried. Now, as this is written, India is back again in the free world.

The Russian strategists have also, over the years, witnessed the defection of Albania (which has again defected, this time *from*

[2] *Honorable Men: My Life in the CIA* (New York: Simon and Schuster, 1978).

China) and, of course, of Yugoslavia and to some extent of Rumania. Only Poland, Bulgaria and East Germany are imagined to be absolutely firm, and, in the case of the Poles, it is hard to suppose that only Zbig Brzezinski is ethnically at odds with the old Russian rulers. A friend of mine holds that our best strategic minds are merely generous men. Seeing the Russian failures in the Third World, they seek to intervene in a kindly way to forestall error and save the Soviets from further trouble. This I doubt.

All joking aside, though it is hard not to have fun, the interventionist records of the United States and the USSR in the former colonial world have been, literally, an unmitigated disaster.

It is one of the prime tenets of the strategic mind that ideological affiliation overrides nationalist sentiment or passion. We can be certain that it does not. And the poorer the country, the greater can be that certainty. That is because nationalist feeling does not diminish with income, but ideological passion does diminish in the absence of capitalism and thus in the absence of anything to socialize or anything much to protect from socialism. Though the point does not figure in modern strategic thought in the East or West, it was basic for Karl Marx.

Also where there is no effective industrial and governmental structure, influence is over a vacuum, which is another thought that seems largely to have escaped the exponents of global strategy and balance. Instead, influence is, at most, over individual politicians who can change their affiliation from one day to the next or be changed from one day to the next. Also, to control only one politician can be to alienate the alternatives and the relevant public, as our Vietnam experience so well revealed. It follows that seeking influence can be a superb design for ensuring that one's influence will come abruptly to an end when the favored politician comes to his end, as later, or more often sooner, he does.

William Colby, though he mentions it as an easily rectified difficulty, is also impressed by the trouble American strategic operatives in Saigon had in communicating with the Vietnamese, with the resulting conflict and misunderstanding. But far more serious is the kind of people who get selected and select themselves for this line of work. Colby, Stockwell and another CIA-affiliated author, Frank Snepp,[3] are all, intentionally or otherwise, emphatic on this point.

In the last century, colonial influence and power were expressed through pedestrian civil servants — in the British case, hard-working, disciplined, unspectacular men who lived within a fairly firm legal frame. In contrast, the sword arm of relentless global strategy is a romantic egoist in whom bravery or bravado — Colby calls it *macho* — takes the place of good sense. The most vivid characterization is in Snepp's book. The CIA station chief in Saigon was not merely lacking in common sense; he had lost touch with reality and lived in a world manufactured out of, and peopled by, his own illusions. Snepp's portrait of the men who had been guiding the Vietnamese to democracy drinking their turn to the helicopters is also unpleasant. However, the older orthodox colonial ventures, virtually without exception, were equally notable for their messy end.

Stockwell is also vivid on the American operatives and even more on the local talent with whom they worked. The station chief in Zaire who directed the Angola intervention in 1975 and 1976 was an ego-hedonist of startling inadequacy. The money that the CIA sent over to hire mercenaries or win support in Zaire went into a series of larcenous rat holes, which, perhaps, was the one thing better than having it used to sustain combat. In the past I've known quite a few CIA people; most of them were sensible, restrained and undramatic men and women engaged in routine reporting and research. I would like to think that the serious and careful public servant is still the

3 *Decent Interval: An Insider's Account of Saigon's Indecent End* (New York: Random House, 1977).

norm. But this, it is terribly clear, isn't the kind of person who is attracted to the strategic and mostly covert enterprises.

It could be supposed that Stockwell and Snepp, since both published their books without official authorization, were inclined to give an unduly unfavorable picture of these CIA troops. Otherwise they would have gone through channels. Unfortunately for this theory, the worst view of all is given by William Colby. He shows, without quite intending it, that irrationality and dementia in covert and strategic operations also occurred at the top. Were the men of strategic mind as effectively self-protective in suppressing books as in ignoring experience, they would have buried William Colby's and let the others go.

Colby's case for irrationality is made partly by exclusion. He never asks why we are competing with the Chinese or the Soviets in Vietnam or elsewhere. Or what we accomplish. The competition itself is the only and sufficient thing. Find a Communist and automatically you react. He also powerfully reinforces my earlier point about national leaders. No matter how unpopular these may be, Colby is unswerving in their support. At the same time (since, to his credit, he would rather be inconsistent than dishonest), in talking of Vietnam he concedes the unpopularity of the ghastly Nhus and affirms his strong belief that the war had to be won with the people. This isn't the only inconsistency that he refuses to paper over. We lost, in his view, because we didn't take "into account the determination of the Vietnamese, Southern as well as Northern, to make their own decisions and fight each other to decide what sort of life Vietnam should lead." But, according to Colby, we were winning almost up to the day of the helicopters. Then Congress cut back on military aid.

Colby is, however, most devastating on the kind of people who carry out the policy. The illicit authors tell of romantic incompetence at the bottom; Colby shows that irresponsibility went right to the top and that misjudgment and the resulting proven capacity for disaster nowise detracted from reputation.

Few men ever so admirably proved themselves incapable of prudent thought as Allen Dulles. In less than a year he was responsible both for the Gary Powers flight over the Paris Summit and the fiasco at the Bay of Pigs. He was then sacked for his errors but wrote a book on the craft of intelligence for which, it appears, he recruited E. Howard Hunt as the expert ghost. Colby remembers Dulles as a "superlative spymaster" and ends his book with a biblical tribute to his insight. Similarly Richard Bissell, who was immediately in charge of the Bay of Pigs (and was in earlier times a very competent economist), emerges from that disaster in Colby's view as a "brilliant, intense" man. All with experience of the CIA knew the late Desmond FitzGerald. He was charming, uncontrollably activist and given dangerously to self-dramatization. When he came to India, I was deeply uneasy and, out of considerable bureaucratic experience, made sure that I knew everything he was doing. Here is Colby on FitzGerald: "I urged that Des FitzGerald be chosen [to be responsible, for God's sake, for all covert activities], but I cautioned Helms that he would have to maintain tight control over him to keep him from charging off into some new Bay of Pigs. Helms did choose FitzGerald . . ." Colby, the source of this remarkable personnel judgment — the selection of the man who would have the single most alarming job in the United States government — went on, himself, to be head of the whole works.

It was the men of relentless strategic mind who guided Lyndon Johnson on Indochina and sent him back to Texas. It was their operatives who went in to bug the Democratic National Committee and sent Richard Nixon back to San Clemente. (Colby disavows Howard Hunt, although he couldn't possibly have been more dangerous than FitzGerald.) I don't especially believe in the rule of three. But I would urge Jimmy Carter to watch those relentless strategic minds. I, personally, would feel far, far safer with Bert Lance.

29

John Dean, Ambition
and the White House

ONE DAY during the 1976 political campaign I flew down from Boston to New York to get a literary award of pleasant intent but negligible intrinsic value. A big crowd was waiting to get on the shuttle, and at the luncheon I told a small fable of how I made it aboard: just ahead of me in line was John Dean, and when he got on, four ex-Nixon hands of depraved appearance hastily got off. No one laughed at my small fantasy, and had I then read Dean's book,[1] I wouldn't have thought it funny either.

John Dean, as all now know, is a man of total recall, and this is combined with a superb instinct for whatever is most damaging to his fellow malefactors. At the time he wrote his book, it was hard to imagine that anyone could come up with anything that would be newly detrimental to Richard Nixon. But Dean has Nixon talking about typewriters and remembering to Chuck Colson that "We built one in the Hiss case," which, of course, is what the Hiss partisans have always most wanted to believe. If true, which I doubt, it robs Dick of the accomplishment on which his whole reprehensible career was founded. (I imagine that, in fact, he mumbled something about building the legal case on the typewriter.) John Connally is also a hard man to

[1] *Blind Ambition: The White House Years* (New York: Simon and Schuster, 1976).

hurt, but Dean has Richard Kleindienst (as Attorney General) exclaiming, "For Christ's sake, John Connally was over here not long ago trying to get me to handle a problem for him. And when I refused, the President started climbing all over my back." What was that one, John?

Then, almost without effort, he ropes in Kleindienst, a law officer who seems not to have supposed that he was meant to report crime, only to complain that, as Attorney General, guilty knowledge of wrongdoing was embarrassing to him. After his departure from the Department of Justice, in a slightly tense ceremony in the favoring presence of one of his successors, Edward Levi, Kleindienst's portrait was unveiled to shine forth there with those of his predecessors. There is enough in Dean's book on the Kleindienst view of the law to require lawyers in the DOJ or elsewhere, if they have any professional pride at all, to get it moved into some reasonably roomy broom closet.

Henry Petersen becomes an excessively reliable conduit to Nixon; Howard Baker, as Sam Dash has also told, is shown to have brought the cover-up into the Ervin Committee while looking like a novitiate scoutmaster on television; a federal judge, Charles Richey, is "a Nixon appointee, who was sending encouraging signals through our contacts."

Even the Christian credentials of Charles Colson take a beating, and it will be a relief to Jesus to have it publicly known that their association is not nearly so intimate as Chuck has been claiming. Here is Colson on Jeb Magruder, long after being born again: "Magruder's full of shit. That bastard tests my Christian patience to the breaking point." Not the language you would expect to hear along the Stations of the Cross.

I greatly enjoyed John Dean's book despite my guilty knowledge as to the reason: the attraction of the Watergate literature lies in your ability to believe that you are reading history while, in fact, you are getting a mean and unworthy pleasure from the disasters of people you don't know but know you don't like.

(Though I've been moderately around, I have never met any of the Watergate delinquents.)

However, in the Dean story there is some redeeming social content, and it could be considerable. He (or more likely a retarded editor) came up with the title *Blind Ambition*; in fact, Dean saw quite clearly that ambition had brought him into the most lethal single location in the United States. The lesson is here.

Although it was on the way before, the White House personnel underwent a huge expansion in the Eisenhower years. In the military tradition, Ike was a man for staff. Perhaps in consequence, though there had been minor misfortunes before, the first major White House casualty was his chief assistant, Sherman Adams. Since then the toll has been great. The Kennedy-Johnson foreign-policy people were deeply damaged. The Nixon domestic-policy people were devastated. The reason lies in the most aberrant act of administrative insanity known to modern government: every four or eight years a large band of men, mostly without previous experience of government, often young, all dangerously euphoric because of recent and sometimes accidental political success, all billed initially as geniuses by the Washington press corps and believing their own notices, all persuaded that they were meant by the stars to reinvent the wheel, are given great ostensible and even actual power in the White House.

Added to the insanity of this action is the absence of any need for congressional clearance, any need to answer to congressional inquiry, an "in" atmosphere that limits conversation to those who are also in and the knowledge that people elsewhere in the government will snap to attention by the phone when word comes that the White House is on the line, even if it is only to ask for the youth unemployment figures. It is also part of the ghastly mystique that the President — variously the leader of the free world, occupant of the world's most powerful post or "the man in the Oval Office" — is entitled to unquestioning fealty even when unquestionably wrong and that, above all, he must be protected, including from those who might usefully tell him that he *is* wrong.

A new Cabinet officer is subject to the formidable discipline of the civil service, which is a rich and often overpowering source of information on the perils of the kind of brilliant innovation which so many before have found fatal. The White House staff is not subject to this restraint. And there is the seductive effect of the royal standard of presidential living — the planes, limousines, special telephones, air-to-ground communications, the offices, mess, doctors and the wonderful ethic of command and obedience — all of which become available in greater or less measure to the denizens. George Reedy was eloquent on all this a few years back. Dean has many words to add, including a small but revealing comment to his lawyer about those who leave the White House: "No matter what they say, it always rips them up. They come back begging for mess privileges and invitations and stuff like that."

For Nixon, we now know, the presidency was overwhelmingly important as compensation for the economic deprivation for which his family had seemed by nature to be intended. Thus the presidential palaces in the sun, the new and more extravagant airplanes of which Dean tells admiringly, the King Carol uniforms and the inauguration extravaganza which Nixon once said is what the presidency is all about. His staff, in consequent imitation, was more than normally concerned with perquisites; Dean tells, with surprise, how Haldeman used an automobile to commute from his quarters to San Clemente, when going by car, helicopter and then car again took only a little longer. I spent the early weeks of the Kennedy Administration in the White House; neither the President nor his entourage was that fascinated by apparatus. But it was a heady experience. We all felt that the country had been forced to wait too long for our attention. And in later years several of my friends (and President Johnson as well) succumbed to their surroundings and the feeling not that they were above the law but that they were above military and associated political error.

As I write this, Governor Carter is preparing to bring a new bunch into the White House. I hope they and he are aware of the terrible risk on which they are entering and that the country is duly alarmed. Mr. Carter does promise a reorganization of the government. On the basis of all recent experience, it should begin with the creation of a nonpolitical White House secretariat, a small group of bureaucrats, no less, who know the perils of the post. He should then ask for legislation limiting to a couple of dozen at most the number of new professional appointments a President can make to the White House staff. Tasks that these men cannot perform should be passed down to the departments and agencies. Perhaps he should ask that the most senior of his staff be subject to Senate confirmation. The Senate will normally be compliant; the faceless men will not thereafter be quite so faceless, and they will have an indication from the beginning that the people and the Congress, not the President, are the ultimate authority.

30

RN:
The Memoirs of Richard Nixon[1]

THE AUTHOR will surely think it appropriate, perhaps inevitable, that I review his book, for he tells, at around his average level of truth, that I both started him on his public career and helped, however marginally, to bring that career to an end.

The beginning was in January of 1942 when he was brought into the Office of Price Administration to work on rubber and tire rationing, tasks then under my direction. (The management of price control and that of rationing were shortly afterward separated, to the benefit of both.) He tells in his book that he had earlier tried for the FBI but was turned down, not on grounds of character but because of an appropriations cut. So once, under Hoover, the FBI had its appropriations cut. However, in line with a notable Nixon tradition, I must plead that I didn't know what was going on. I didn't meet Mr. Nixon either then or later.

My contribution to ending his career was in the mid-sixties when the Kennedy Library asked me for my papers. On leaving OPA, the State and War Departments and on ceasing to be an ambassador, I didn't think to take my official papers, which I supposed belonged to the government and, in any case, were

[1] (New York: Grosset & Dunlap, 1978.)

rather bulky. (I've since had to go to the National Archives on occasion to look things up.) I did send to the Kennedy Library manuscripts of *The Affluent Society, The Great Crash*, other folk classics, a mass of personal correspondence and, by accident, our canceled checks, bank statements, marriage license, my naturalization certificate and our old income tax returns. Taking a tax deduction on old tax returns would have been a major breakthrough in sophisticated tax avoidance. However, David Powers, the Kennedy friend and librarian, sent these, the checks and the other detritus back.

On the manuscripts and correspondence, including that with J.F.K., his wife, Adlai Stevenson, Averell Harriman, Arthur Schlesinger, numerous editors, economists and literary types, Nathan Marsh Pusey and various government loyalty boards, the Internal Revenue Service allowed me a stunning (as it then seemed) $4500. This, Mr. Nixon here says, later encouraged him to take a deduction of $576,000 on his more official papers, with endlessly damaging results, made worse because that kind of deduction had ceased to be legal and the gift was backdated without his knowledge. He also blames Lyndon Johnson, George Wallace, Hubert Humphrey and others for putting him on this downward path.

I tell the foregoing partly for my own pleasure but also because it reflects a central theme of this book. Mr. Nixon never did anything wrong unless someone else had done it first. And all evil disappears if it has a precedent.

As committee work goes, his book is not badly written. And, as in the matter just cited, it bears the undoubted imprint, for better or worse, of Mr. Nixon's personality, which is certainly not without interest. He is imaginative and resourceful, and, if one is to believe in democracy, one must find some qualities to explain and justify his twice serving as Vice-President, twice being elected President. It isn't as though he weren't known. I think it was this resourcefulness and imagination that allowed him to see, along with Henry Kissinger, the opportunity that the frozen Cold War diplomacy of his predecessors had given him. Thus the ensuing and important improvement in relations with China and the Soviet Union. This resourcefulness also explains his almost incredible ability (until Watergate) to battle his way out of an endless series of personal and political disasters — misfortunes, it is even more interesting to reflect, which were invariably invited by his own bad judgment.

In turn, the source of this bad judgment, one gathers, was his inability to think, or anyhow believe, that he himself could do anything wrong. That Nixon was a rascal is now generally accepted. But, as his political career revealed and this book superbly affirms, he was and remains a rascal who either considers himself a deeply moral man or, more precisely, believes that he can so persuade any known audience.

This belief bears heavily on the reader of this book. Were it a candid account of the author's lifetime of political and personal tergiversation, it would be wonderful reading. But the compulsive search for precedent, the innumerable explanations of the ineffable and the belief that all his accusers are ill-motivated except when they say something on his side (as even Archibald Cox once did) get to be a bore.

Not entirely, however, for one gets caught up in the game — in looking for the flaw in his explanation. Thus, as he sees it, one of the typically egregious misunderstandings to which he was subject concerned that $100,000 that Bebe Rebozo got in 1970 from Howard Hughes. Bebe, "one of the kindest and most generous men I have ever known," intended to hold it for the 1972 campaign. But then various feuds within the Hughes empire and the danger of exposure and embarrassment caused him to sock it away in a safe deposit box and not even tell Nixon about it for a long while. Later, when the risk of exposure was still acute, it was returned. There couldn't have been anything funny about the transaction, according to Nixon, for an inspection of the bills showed that they had been issued by the Treasury well before the date when Rebozo said he got the money. No switch.

One's mind dwells, inevitably, on why the payment was in cash, why 1972 campaign contributions were being collected so early, what would have happened to the bills had there been no danger of exposure. Here is the real interest. Nixon's explanations almost always invite such thoughts, and they don't greatly tax the mind. I'm persuaded that most of his disasters came from his conviction that if he could persuade himself that something was virtuous or legitimate, he could persuade almost everybody else.

There are other aspects of the Nixon personality that command attention. Sometimes it is less the resourcefulness than the convolution:

> I was concerned that Haldeman handle the matter deftly. I did not want him to strong-arm Helms and Walters, nor did I want him to lie and say there was no involvement. I wanted him to set out the situation in such a way that Helms and Walters would take the initiative and go to the FBI on their own. I told Haldeman to say that I believed this thing would open up the whole Bay of Pigs matter — to say that the whole thing was a sort of comedy of errors and that they should call the FBI in and say that for the sake of the country they should go no further into this case.

This passage makes one weep even for Haldeman. Lompoc or whatever must have been a blessed relief.

Nixon is not only a master of convoluted thought, but he appreciates it in others. Here is his account of his briefing by Dean Rusk on Vietnam during the 1968 campaign:

> Rusk, one of the ablest and most honorable men ever to serve as Secretary of State, made the point that the rest of Asia would be in a "panic" if the United States were to withdraw from Vietnam without an honorable peace settlement. He said that he held this view completely apart from the domino theory, which he considered simplistic. He believed that American withdrawal from Vietnam would leave the Chinese Communists as the only major power on the Asian mainland, thus creating the panic.

The ability to distinguish between the simplistic domino theory and the sophisticated panic theory is the kind of thing Mr. Nixon cherishes. He also responds well to a really complex metaphor. He remembers and quotes Bryce Harlow as telling him on April 23, 1973, that "If Haldeman, Dean and Ehrlichman have undertaken actions which will not float in the public domain, they must leave quickly — they are like a big barnacle on the ship of state, and there is too much at stake to hang on for personal reasons."

Perhaps because of professional pecuniary interest, I'm against people boycotting books. And quite a few will find the reading and associated unraveling, if not a pleasure, at least a challenge. I'm also against a too violent reaction to Mr. Nixon's belief, here affirmed, that the misuse of the FBI, the IRS and other federal agencies is one of the accepted rights of incumbency. He tells with sorrow that Patrick Gray and the FBI were out of his control in their misguided pursuit of wrongdoing and that the IRS was not adequately motivated in the investigation he ordered of Larry O'Brien and the McGovern supporters during the 1972 campaign. These efforts *were* indefensible, but the book is a testament to the number in the federal government who would not go along with any such wrongdoing.

Indeed, next only to the self-justification and the convolution, the book comes across as Mr. Nixon's bill of complaint against civil servants, members of Congress, Kennedy types, newspapers and television people who were out to sabotage his intentions. These intentions were bad, but Nixon was far from effecting them. Our liberties were not curtailed. On the contrary, there never was a time when more Americans were expressing themselves more stridently and diversely with so little fear as in the Nixon years. I'm firm in the belief, a minority view of long standing, that anyone who spoke out under Richard Nixon had little to fear, anyone who was shut up by Richard Nixon had nothing to say.

Appendix
Index

Power and the Useful Economist[1]

WITHIN THE LAST DOZEN YEARS what before was simply known as economics in the nonsocialist world has come to be called neoclassical economics. Sometimes, in tribute to Keynes's design for government intervention to sustain purchasing power and employment, the reference is to Keynesian or neo-Keynesian economics. From being a general and accepted theory of economic behavior, this has become a special and debatable interpretation of such behavior. For a new and notably articulate generation of economists a reference to neoclassical economics has become markedly pejorative. In the world at large the reputation of economists of more mature years is sadly in decline.

However, the established economics has reserves of strength. It sustains much minor refinement which does not raise the question of overall validity or usefulness and which is agreeable employment. It survives especially in the textbooks, although even in this stronghold one senses anxiety among the more progressive or commercially sensitive authors. Perhaps, they are asking, there are limits to what the young will accept.

The arrangements by which orthodoxy is conserved in the modern academy are also still formidable. Economic instruction in the United States is about a hundred years old. In its first half century economists were subject to censorship by outsiders.

[1] This was my presidential address at the Eighty-fifth Annual Meeting of the American Economic Association in Toronto, Canada, in December 1972.

Businessmen and their political and ideological acolytes kept watch on departments of economics and reacted promptly to heresy, the latter being anything that seemed to threaten the sanctity of property, profits, a proper tariff policy and a balanced budget, or that suggested sympathy for unions, public ownership, public regulation or, in any organized way, for the poor. The growing power and self-confidence of the educational estate, the formidable and growing complexity of the subject matter of economics and, no doubt, the increasing acceptability of our ideas to conservatives have largely relieved us of this intervention. In leading centers of instruction faculty government is either secure or increasingly so. But in place of the old censorship has come a new despotism. This consists in defining scientific excellence in economics not as what is true but as whatever is closest in belief and method to the scholarly tendency of the people who already have tenure in the subject. This is a pervasive test, not the less oppressive for being, in the frequent case, both self-righteous and unconscious. It helps ensure, needless to say, the perpetuation of the neoclassical orthodoxy.

There are, however, problems even with this control. Neoclassical and neo-Keynesian economics, though providing unlimited opportunity for the demanding niceties of refinement, has a decisive flaw. It offers no useful handle for grasping the economic problems that now beset the modern society. And these problems are obtrusive; they will not lie down and disappear as a favor to our profession. No arrangement for the perpetuation of ideas is secure if the ideas do not make useful contact with the problems they are presumed to illuminate or resolve.

I propose in this essay to mention the failures of neoclassical economics. But I want also to urge the means by which we can reassociate ourselves with reality. Some of this will summarize arguments I have made at greater length on other occasions. Here even conservatives will be reassured. To adumbrate and praise one's own work is in the oldest and most reputable tradition of our profession.

The most damaging feature of neoclassical and neo-Keynesian economics is the arrangement by which power — the ability of persons or institutions to bend others to their purposes — is removed from the subject. The business firm is said to be wholly subordinate to the instruction of the market and thereby to the individual or household. The state is similarly subordinate to the instruction of the citizen. There are exceptions, but they are to the general and controlling rules, and it is firmly on these rules that neoclassical theory is positioned. If the business firm is subordinate to the market — if that is its master — then it does not have power to deploy in the economy, save as this is in the service of the market and the consumer. And it cannot bring power to bear on the state, for there the citizen is fully in charge.

The decisive weakness is not in the assumptions by which neoclassical and neo-Keynesian economics elides the problem of power. The capacity even (perhaps especially) of scholars for sophisticated but erroneous belief based on conventionally selected assumptions is very great, particularly where this coincides with convenience. Rather, in eliding power — in making economics a nonpolitical subject — neoclassical theory destroys the relation of economics to the real world. In that world, power is decisive in what happens. And the problems of that world are increasing both in number and in the depth of their social affliction. In consequence, neoclassical and neo-Keynesian economics relegates its players to the social sidelines. They either call no plays or urge the wrong ones. To change the metaphor, they manipulate levers to which no machinery is attached.

Specifically, the exclusion of power and the resulting political content from economics causes it to identify only two intrinsic and important faults in the modern economy. One of these is the problem of market imperfection — more specifically, of monopoly or oligopoly, control of a market by one firm or jointly by a few. This fault leads, in turn, to insufficient investment and output and to unnecessarily high prices. The other fault is a tendency to unemployment or inflation — to a deficiency or excess in the aggregate demand for goods and services in the economic system as a whole. The remedies to which the accepted economics then proceeds are either ridiculous, wrong or partly irrelevant. Neither its microeconomic nor its macroeconomic policy really works.[2] Meanwhile it leaves other urgent economic issues untouched and mostly unmentioned. Let me specify.

Beginning with monopoly and oligopoly, it is now the considered sense of the community, even of economists when unhampered by professional doctrine, that the most prominent areas of market concentration — automobiles, rubber, chemicals, plastics, alcohol, tobacco, detergents, cosmetics, computers, bogus and other health remedies, space adventure — are areas not of low but of high development, not of inadequate but, more likely, of excessive resource use. And there is a powerful instinct that in some areas of monopoly or oligopoly, most notably in the production of weapons and weapons systems, resource use is dangerously vast.

In further contradiction of the established conclusions, there is much complaint about the performance of those industries where market concentration is the least — the industries which, in number and size of firms, most closely approach the neoclassical market ideal. Housing, health services and, potentially, food supply are the leading cases. The deprivation and social distress

---

[2] Neoclassical economics, in modern times, has divided itself into two broad areas of specialization, research and instruction. There is microeconomics, which concerns itself with firms, industries and their response to the market. And there is macroeconomics, which involves itself with aggregative movements in the economy — with Gross National Product and National Income and with employment and general price movements.

that follow from the poor performance of these industries nearly all economists, when not in the classroom, take for granted.

The well-blinkered defender of established doctrine argues that the ample resource use in the monopolistic industries and the deprivation in the dispersed small-scale industries reflect the overriding fact of consumer preference and choice. And in the areas of deprivation he insists that the fault lies with firms that, though small, are local monopolies or reflect the monopoly power of unions. These explanations beg two remarkably obvious questions. Why does the modern consumer increasingly insist on self-abuse and increasingly complain of the discomforts from that self-assault? And why do the little monopolies perform so badly and the big ones so well?

In fact, neoclassical economics has no explanation of the most important microeconomic problem of our time. That is why we have a highly unequal development as between industries of great market power and industries of slight market power, with the development, in defiance of all doctrine, greatly favoring the first.

The failure in respect to unemployment and inflation has been, if anything, more embarrassing. Save in its strictly mystical manifestation in one branch of monetary theory, the accepted policy on these matters depends for its vitality and workability on the existence of the neoclassical market. That market, whether competitive, monopolistic or oligopolistic, must be the ultimate and authoritative instruction to the profit-maximizing firm. When output and accompanying employment are deficient, the accepted policy requires that aggregate demand be increased; this increases market demand, and to this firms, in turn, respond. They increase output, add to their labor force and reduce unemployment.

When output in the economy is at or near the effective capacity of plant and labor force, prices rise and inflation becomes the relevant social discomfort. The remedy is then reversed. Demand is curtailed, and the result is either an initial effect on prices or a delayed one as surplus labor seeks employment, interest rates fall and lower wage and material costs bring stable or lower prices.

Such is the accepted basis of the policy. It follows fully from the neoclassical faith in the market. The market renders its instruction to the producing firm. The latter cannot, because of competition, much raise its prices while there is idle capacity and unemployment. When these disappear, restraints on demand through monetary or fiscal policy or some combination of the two can prevent it from doing so. The practical consequences from pursuing this policy need no elucidation. It has been tried in recent years in every developed country. The result has been either politically unacceptable unemployment or persistent and socially damaging inflation or, normally, a combination of the two. That combination the neoclassical system does not and cannot contemplate. Modern medicine would not be more out of touch with its world if it could not embrace the existence of the common cold.

We should not deny ourselves either the instruction or the amusement that comes from the recent history of the United States in this matter. In 1969, Mr. Nixon came to office with a firm commitment to neoclassical orthodoxy. Any direct interference with wages or prices was explicitly condemned. In this position he was supported by some of the most distinguished and devout exponents of neoclassical economics in all the land. His later announcement that he was a Keynesian involved no precipitate or radical departure from this faith; the discovery came nearly thirty-five years after *The General Theory*.[3] But in 1971, facing reelection, Mr. Nixon found that his economists' commitment to neoclassical and neo-Keynesian orthodoxy, however admirable in the abstract, was a luxury he could no longer afford. He apostatized to wage and price control; so, with exemplary flexibility of mind, did his economists.

---

[3] John Maynard Keynes, *The General Theory of Employment Interest and Money* (New York: Harcourt Brace, 1936).

There was an effort to reconcile the need for controls with the neoclassical market. This involved an unrewarding combination of economics and archeology with wishful thinking. It held that an inflationary momentum developed during the late nineteen-sixties in connection with the financing — or underfinancing — of the Vietnam war. And inflationary expectation became part of business and trade union calculation. The momentum and expectation survived. The controls would be necessary until the inflationary momentum was dissipated. Then the neoclassical and neo-Keynesian world would return, along with the appropriate policies in all their quiet comfort: no inflation; no serious unemployment. We may be sure that will not happen. Nor will we expect it to happen if we see the role of power and political decision in modern economic behavior.[4]

The assumptions that sustain the neoclassical and neo-Keynesian orthodoxy can no longer themselves be sustained. The growth of the modern great corporation has destroyed their validity. Instead of the widely dispersed, essentially powerless firms of neoclassical orthodoxy, we must now come to terms with the world of the modern large corporation. Laymen will be astonished that we have not already done so.

Specifically, we must accept that for around half of all economic output there is no longer a market system but a power or planning system. (Power is used to control what was previously external to the firm and thus unplanned. To stress not the instrument, power, but the process, planning, seems to be more descriptive as well as, possibly, less pejorative.) The plan-

---

4 As I've said, this was written in the autumn of 1972 for the Christmas holiday meetings of the American Economic Association. Early in 1973, in accordance with the doctrine just adumbrated, Mr. Nixon's economists urged and obtained the abandonment of controls. They promised price stability; there followed the worst peacetime inflation so far in our history. This was eventually arrested, though not completely, by the most serious recession and the most severe unemployment since the Great Depression. There is danger in praising one's foresight, for the gods are not always kind. But the risk on these matters is less than usual.

ning system in the United States consists of, at the most, 2000 large corporations. They do not simply accept the instruction of the market. Instead they have extensive power over prices and also over consumer behavior. They rival, where they do not borrow from, the power of the state. My conclusions on these matters will be somewhat familiar, and I shall spare myself the pleasure of extensive repetition. The power that these ideas ascribe to the modern corporation in relation to both the market and the state, the purposes for which it is used and the associated power of the modern union would not seem implausible or even very novel were they not in conflict with the vested doctrine.

Thus we agree that the modern corporation, either by itself or in conjunction with others, has extensive influence over its prices and often over its major costs. And, accepting the evidence of our eyes and ears, we know that it goes beyond its prices and the market to persuade its customers. We know also that it goes back of its costs to organize supply. And it is commonplace that from its earnings or the possession of financial affiliates it seeks to ensure and control its sources of capital. And likewise that its persuasion of the consumer, joined with the similar effort of other firms — and with the more than incidental blessing of neoclassical pedagogy — helps establish the values of the community, notably the association between well-being and the continuously increased consumption of the products of this part of the economy.

As citizens if not as scholars, we further agree that the modern corporation has a compelling position in the modern state. What it needs in research and development, technically qualified people, public works, emergency financial support when troubles loom, socialism when profit ceases to be probable, becomes public policy. So does the military procurement that sustains the demand for numerous of its products. So, perhaps, does the foreign policy that justifies the military procurement. And the means by which this power is brought to bear on the state is widely accepted. It requires an organization to deal with an organization, and between public and private bureaucracies — between General Dynamics and the Pentagon, General Motors and the Department of Transportation — there is a deeply symbiotic relationship. Each of these organizations can do much for the other. There has been between them a large and continuous interchange of executive personnel.

Finally, over this exercise of power and much enhancing it is the rich gloss of reputability. The men who guide the modern corporation and the outlying financial, legal, legislative, technical, advertising and other sacerdotal services of corporate function are the most respectable, affluent and prestigious members of the national community. They are the Establishment. Their interest tends to become the public interest. It is an interest that numerous economists find it comfortable and rewarding to avow, while denying in instruction and thought the power that produces that reward.

The corporate interest is profoundly concerned with power — with winning the acceptance by others of the collective or corporate purpose. This interest includes the profits of the firm. These are a measure of success. They also ensure the freedom of the management — what I have called the technostructure — from stockholder interference. The stockholders in the large corporation are aroused, if at all, only by inadequate earnings. And profits are important because they bring the supply of capital within the control of the firm. But of greater importance is the more directly political goal of growth. Growth carries a specific economic reward; it enhances the pay, perquisites and opportunities for promotion of the members of the technostructure, and it rewards most those whose product or service is growing most. But growth also consolidates and enhances authority. It does this for the individual — for the man who now heads a larger organization or a larger part of an organization than before. And it increases the influence of the corporation as a whole.

The unmanaged sovereignty of the consumer, the ultimate sovereignty of the voter and the maximization of profits with the resulting subordination of the firm to the market are the three legs of a tripod on which the accepted neoclassical system stands. These are what exclude the role of power in the system. All three propositions, it will be seen, tax the capacity for belief. That the modern consumer is the object of a massive management effort by the producer is not readily denied. The methods of such management, as noted, are embarrassingly visible. Modern elections are fought extensively on the issue of the subordination of the state to corporate interest. As voters, economists accept the validity of that issue. Only their teaching denies it. But the commitment of the modern corporate bureaucracy to its expansion is, perhaps, the clearest of all. That the modern conglomerate pursues profit over aggrandizement is believed by none. It is a commonplace of these last years, strongly reflected in the price of securities, that agglomeration has been good for growth, bad for earnings.

There does remain in the modern economy — and this I stress — a world of small firms where the instruction of the market is still paramount, where costs are given, where the state is remote and subject through the legislature to the traditional pressures of economic interest groups and where profit maximization alone is consistent with survival. We should not think of this as the classically competitive part of the system, in contrast with the monopolistic or oligopolistic sector from which the planning system has evolved. Rather, in its admixture of competitive and monopolistic structures, it approaches the neoclassical model. The corporation did not take over part of the neoclassical system. It moved in beside what the textbooks teach. In consequence, we have the two systems. In one, the power of the firm is still, as ever, contained by the market. In the second and still evolving system, this power extends incompletely but comprehensively over markets, over the people who patronize them, over the state and thus, ultimately, over resource use. The coexistence of these two systems becomes, in turn, a major clue to economic performance.

Power being so comprehensively deployed in a very large part of the total economy, there can no longer, except for reasons of game-playing, busy work or more deliberate intellectual evasion, be any separation between economics and politics. When the modern corporation acquires power over markets, power in the community, power over the state and power over belief, it is a political instrument, different in form and degree but not in kind from the state itself. To hold otherwise — to deny the political character of the modern corporation — is not merely to avoid the reality. It is to disguise the reality. The victims of that disguise are the students we instruct in error. The beneficiaries are the institutions whose power we so disguise. Let there be no question: economics, so long as it is thus taught, becomes, however unconsciously, a part of the arrangement by which the citizen or student is kept from seeing how he is, or will be, governed.

This does not mean that economics now becomes a branch of political science. Political science is the captive of the same stereotype — the stereotype that the citizen is in effective control of the state. Political science too must come to terms with corporate enterprise. Also, while economics often cherishes thought, at least in principle, political science regularly accords reverence to the man who knows only what has been done before. Economics does not become a part of political science. But politics does become a part of economics.

There will be fear that once we abandon the present theory with its intellectually demanding refinement and its increasing instinct for measurement, we shall lose the filter by which scholars are separated from charlatans and windbags. The latter are a danger, but there is more danger in remaining with a world that is not real. And we shall be surprised, I think, at the new clarity and intellectual consistency with which we see our world once power is made a part of our system. To such a view let me now turn.

In the neoclassical view of the economy a general identity of interest between the goals of the business firm and those of the community could be assumed. The firm was subject to the instruction of the community through either the market or the ballot box. People could not be fundamentally in conflict with themselves. However, once the firm in the planning system is seen to have comprehensive power to pursue its own interest, this assumption becomes untenable. Perhaps by accident its interests are those of the public, but there is no organic reason why this must be so. In the absence of proof to the contrary, divergence of interest between individual and corporation, not identity of interest, must be assumed.

The nature of the conflict also becomes predictable. Growth being a principal goal of the planning system, it will be the greatest where power is greatest. In the market sector of the economy, growth will, at least by comparison, be deficient. This will not be, as neoclassical doctrine holds, because people have a congenital tendency to misunderstand their needs. It will be because the system is so constructed as to serve their needs badly and then to win greater or less acquiescence in the result. That the present system should lead to an excessive output of automobiles, an improbable effort to cover the economically developed sections of the planet with asphalt and a fantastically expensive and potentially suicidal investment in missiles, submarines, bombers and aircraft carriers is as one would expect. These are the industries with power to persuade and to command resources for growth. Thus does the introduction of power as a comprehensive aspect of economic thought correct present error. These, however, are exactly the industries in which the neoclassical view of monopoly and oligopoly (and of associated price enhancement and profit maximization at the expense of resource use and production) would, of all things, suggest a controlled inadequacy of output. How wrong are we allowed to be!

The counterpart of relatively too great a share of manpower, materials and investment in the planning system, where power is comprehensively deployed, is a relatively deficient use of such resources where power is absent. Such will be the flaw in the part of the economy where competition and entrepreneurial monopoly, as distinct from great corporate organization, are the rule. And if the product or service so penalized is closely related to comfort or survival, the resulting discontent will be considerable. That housing, health services and local transportation are now areas of grave inadequacy is agreed. It is in such industries that all modern governments seek to expand resource use. Here, in desperation, even devout free-enterprisers accept the need for social action, even socialism.

Economics serves badly this remedial action. Its instruction not only disguises corporate power but makes remedial action in housing, health care and transportation abnormal — the consequence of *sui generis* error that is never explained. What should be seen as a necessary and legitimate function of government appears, instead, as some kind of accident. This is not the mood that conduces to the imagination, pride and determination which should characterize such important public action.

When power is admitted to our calculus, our professional embarrassment over the coexistence of unemployment with inflation also disappears. Economics makes plausible what governments are forced, in practice, to do. Corporations have power in their markets. So, and partly in protective response, do unions. The competitive claims of unions can most conveniently be resolved by passing the cost of settlement along to the public. Measures to arrest this exercise of power by limiting the aggregate demand for goods must be severe. Only if there is much unemployment, much idle capacity, is the ability to raise prices impaired. Until then, unemployment and inflation coexist. Not surprisingly, the power of the planning system has also been used to favor those restraints on demand that have least effect on its operations. Thus monetary policy is greatly favored. This policy operates by restricting bank lending. Its primary effect, in consequence, is on the neoclassical enterpreneur — the construction firm, for example — which does business on borrowed money. It has little impact on the large established corporation which, as an elementary exercise of power, has ensured itself a supply of capital from earnings or financial affiliates. The power of the planning system in the community has also won immunity for public expenditures important to itself — highways, industrial research, rescue loans, national defense. These have the sanction of a higher public purpose. If demand must be curtailed, these are excepted. There has been similar success with corporate and personal taxes. They are what you now reduce to stimulate employment, support the incentive to invest and ensure against capital shortages. In such fashion, fiscal policy has been accommodated to the interests of the planning system. This has been done with the support of economists in whose defense it must be said that they are not usually aware of the forces by which they are moved.

In this view of the economy we see also the role of controls. The interaction of corporate and trade union power can be made to yield only to the strongest fiscal and monetary restraints. Those restraints that are available have a comparatively benign effect on those with power, but they weigh adversely on people who vote. When no election is in prospect, such a policy is possible. It will earn applause for its respectability. But it cannot be tolerated by anyone who weighs wisely its popular effect.

As with the need for social action and organization in the market sector, there are many reasons why it would be well were economists to accept the inevitability of wage and price controls. It would help keep politicians, when responding to the resonance of their own past instruction, from supposing controls to be wicked, unnatural and hence temporary and to be abandoned whenever they seem to be working.[5] This is a poor mood in which to develop sound administration. And it would cause economists themselves to consider how controls can be made workable, how the effect on income distribution can be made equitable. With controls this last becomes a serious matter. The market is no longer a device for legitimizing inequality, however egregious, in income distribution. Much inequality is then seen to be, as it is, the result of relative power.

There are differences in development, in performance, as between the planning and market sectors of the economy. With them goes a difference in income between the two great sectors of the economy. In the neoclassical system it is assumed that labor and capital will move between industries — from lower to higher — to equalize interindustry return. If there is inequality, it is the result of barriers to such movement. Now we see that, given its comprehensive market power, the planning system can protect itself, as a matter of course, from adverse movements affecting its income. The same power allows it to accept unions, for it need not absorb, even temporarily, their demands. In the market system, limited areas of monopoly or union power apart, there is no similar control. And in this sector of the economy, because of the absence of market power, there can be no similar yielding on wage costs, for there is no similar certainty that they can be passed on. It is because of the market character of the industry he sought to organize, not his original power, that Cesar Chavez was for so many for so long the new Lenin. In chemicals or heavy equipment he would not have been noticed. In the market system the self-employed have

---

[5] When Secretary of the Treasury George Schultz announced the abandonment of the Nixon controls a few months after this was written, he said, in effect, that it was because they were working.

the option of reducing their own wages (and sometimes those of families or immediate employees) in order to survive. That possibility does not exist in the highly organized planning system. This is the source of further inequality.

Thus there is a built-in inequality in income between the two systems. And thus also the case in the market sector for minimum wage legislation, support to trade unions in agriculture, price support legislation and, most important perhaps, a floor under family income as antidotes to inequality. All of these measures have their primary impact on the market sector. And again this view of matters fits our present concerns. Minimum wage legislation, price support legislation and support to collective bargaining are questions of continuing political moment as they apply to small business and agriculture. They are not issues in highly organized industry — in the planning system. And the question of a floor under family income, a further matter of political interest, has shown some indication of dividing workers in the planning system, who would not be beneficiaries, from those in the market system, who would be. There is surely some reassurance in a view of the economy that prepares us for the political questions of our time.

The inclusion of power in our economic calculus also brings into focus the debate over the environment. It is the claim of neoclassical economics that it foresaw possible environmental consequences from economic development. It early embraced the notion of external diseconomies of production and, by inference, of consumption. The price of the product did not include the cost of washing out the soot that descended from the factory chimney on the people of the surrounding city. The owner of the automobile or cigarette did not pay for the damage to other people's air and lungs. Alas, this is a modest claim. The noninclusion of external diseconomies was long viewed as a minor defect of the price system — an afterthought, obtaining at most a paragraph in the textbooks or a comment in classroom discussion. And the notion of external diseconomies does not offer a useful remedy. No one can suppose, or does suppose, that more than a fraction of the damage — especially that to the beauty and tranquillity of our surroundings — could be compensated for in any useful way by including in the cost of the product a provision for remedying the damage from its production or use.

If growth is the central and rewarding purpose of the firm and if power is comprehensively available to impose this goal on the society, conflict between the private interest in that growth and the public interest in the environment is inherent. And also inherent, since this power depends extensively not on force but on persuasion, is the effort to make pollution seem palatable or worth the cost, including the effort to make the advertising of remedial action a substitute for action. And so is the remedy to which all industrial countries are being forced. This is not to internalize external diseconomies, add them to the price. It is to specify the legal parameters within which growth may proceed. Or, as in the case of automobile use in central cities, airplane use over urban areas, the SST or industrial, commercial and residential appropriation of countryside and roadside, to prohibit development that is inconsistent with the public interest. We would have saved ourselves much corruption of our surroundings if our economics had held such result to be the predictable consequence of the pursuit of present economic gals and not the exceptional result of a peculiar aberration of the price systems. We see again how the accepted economics supports not the public but the special interest.

Finally, when power becomes part of our system, so does Ralph Nader. If the consumer is the ultimate source of authority, his abuse is an occasional fault. He cannot be fundamentally at odds with an economic system that he commands. But if the producing firm has comprehensive power and purposes of its own, there is every likelihood of conflict. Technology is then subordinate to the strategy of consumer persuasion. Products are changed not to make them better but to take advantage of the belief that what is different is better. There is a high failure rate in engineering because its preoccupation is not with what is good but with what can be sold. So the unpersuaded or disenchanted consumer rebels. This is not a rebellion against minor matters of fraud or misinformation. It is a major reaction against a whole deployment of power by which the consumer is made the instrument of purposes that are not his own.

There are two conclusions to which this exercise — the incorporation of power into our system — compels us. The first is encouraging. It is that economists' work is not yet done. On the contrary, it is just beginning. If we accept the reality of power as part of our system, we have years of useful professional work ahead of us. And since we will be in touch with real issues, and since issues that are real inspire passion, our life will again be pleasantly contentious, perhaps even usefully dangerous. Members of the profession will be saved from the paltry suburban slumber that is the fate of the passive, irrelevant or harmless scholar.

The other conclusion concerns the state. For when we make power and therewith politics a part of our system, we can no longer escape or disguise the contradictory character of the modern state. The state is the prime target in the exercise of economic power. In greater or less measure it is captured by the planning system. Yet on all the matters I have mentioned — organization to offset inadequate performance in such areas as housing and health care, wage and price controls, action to correct systemic inequality, protection of the environment, protection of the consumer — remedial action lies with the state.

Thus perhaps the greatest question of social policy in our time: is the emancipation of the state from the control of the planning system possible?

I would be presumptuous to say yes, even more so to suggest that it will be easy. But there is a gleam of encouragement. Elections are now being fought extensively over issues where the purposes of the planning system diverge from those of the public. The question of defense expenditures is such an issue. That of tax reform is another. The deprivation in housing, mass transportation and health services is yet another — one that reflects the relative inability of these industries to organize and command resources. The question of a guaranteed income, though momentarily quiescent, is another. The environment is such an issue — with its conflict between the technostructure's goal of growth and the public's concern for its surroundings. So is wage and price control. Our politics, forced by circumstance, are coming to accept and deal with this great contradiction between the needs of the planning system and the needs of the public.

It would not be wrong, I believe, to ask that in this effort, economists identify themselves with the public interest, not that of the corporations and the planning system. But if that is too much, I would happily settle for neutrality. Economics now tells the young and susceptible (and also the old and vulnerable) that economic life has no content of power and politics because the firm is safely subordinate to the market and the state and for this reason it is safely at the command of the consumer and citizen. Such an economics is not neutral. It is the influential and invaluable ally of those whose exercise of power depends on an acquiescent public. If the state is the executive committee of the great corporation and the planning system, it is partly because neoclassical economics is its instrument for neutralizing the suspicion that this is so.

*

Index

About the Author

JOHN KENNETH GALBRAITH has achieved distinction in the fields of economics, education, public service, and diplomacy, as well as being an accomplished writer of both fiction and non-fiction. Among his many books are *The Affluent Society, A China Passage, Economics and the Public Purpose, Economics, Peace and Laughter, The Liberal Hour,* and *The New Industrial State,* each published by New American Library.